AI BASED PV POWER G

GEETHAMAHALAKSHMI G

TABLE OF CONTENTS

CHAPTER NO.	TITLE	PAGE NO.

CHAPTER NO.	TITLE	PAGE NO.

CHAPTER NO.	TITLE	PAGE NO.

CHAPTER NO.	TITLE	PAGE NO.

CHAPTER NO.	TITLE	PAGE NO.

LIST OF TABLES

LIST OF FIGURES

FIGURE NO.	TITLE	PAGE NO.

LIST OF SYMBOLS AND ABBREVIATIONS

AHB	-	Adaptive Hysteresis Band
ANFIS	-	Adaptive Neuro-Fuzzy Inference System
AFSA	-	Artificial Fish Swarm Algorithm
AI	-	Artificial Intelligence
AMV	-	Atmospheric Motion Vector
AGC	-	Automatic Generation Control
AMR	-	Automatic Meter Reading
AR	-	Autoregressive
ARMA	-	Auto-Regressive Moving Average
ARX	-	Autoregressive with Exogenous Input
ARMAX	-	Autoregressive-Moving-Average Method with Exogenous Inputs
BPNN	-	Backward Propagation Neural Network
BNN	-	Bayesian Neural Network
CHB	-	Cascaded H-Bridge
CLARA-A2	-	Cloud, Albedo, Radiation dataset Edition 2
COP	-	Coefficients of Performance
CEEMDAN	-	Complete Ensemble Empirical Mode Decomposition Adaptive Noise
CONUS	-	Continental United States
DNN	-	Deep Neural Network
DR	-	Demand Response
DSM	-	Demands Side Management
DE	-	Differential Evolution
DC	-	Direct Current
DG	-	Distributed Generation
DL Model	-	Deep Learning Model
ENN	-	Elman Neural Network

E	-	Error
E_C	-	Error Change(Change of Error)
ECMWF	-	European Centre for Medium Range Weather Forecasts
ESRA	-	European Solar Radiation Atlas
ESSS	-	Exponential-Smoothing -State Space
ELM	-	Extreme Learning Machine
FS	-	Feature Selection
FFNN	-	Feedforward Neural Network
FSPOA	-	Fickle Step-Size Perturb and Observe Algorithm
FFA	-	Firefly Algorithm
FC	-	Flying Capacitor
FL	-	Fuzzy Logic
FLC	-	Fuzzy Logic Controller
FRF-SVM	-	Fuzzy Regression Functions with SVM
GT	-	Gamma Test
GRNN	-	Generalized Regression Neural Networks
GA	-	Genetic Algorithm
GFS	-	Global Forecast System
GHI	-	Global Horizontal Irradiance
GMPP	-	Global MPP
GP	-	Global Peak
GSI	-	Global Solar Irradiance
GCPV	-	Grid Connected Photovoltaic
GPV	-	Grid Point Value
GGA	-	Grouping Genetic Algorithm
HWSECS	-	Hybrid Wind-Solar Energy Conversion System
IMFs	-	Intrinsic Mode Functions
I&T	-	Irradiance and Temperature
KF	-	Kalman Filter
KNN	-	K-Nearest Neighbor

LVQ	-	Learning Vector Quantization
LBC	-	Level Boosting Circuit
LMPP	-	Local MPP
MPPT	-	Maximum Power Point Tracking
MAE	-	Mean Absolute Error
MAPE	-	Mean Absolute Percentage Error
MBE	-	Mean Bias Error
MSE	-	Mean Square Error
MRMR	-	Minimum Redundancy – Maximum Relevance
MFO	-	Moth-Flame Optimization
MA	-	Moving Average
MLI	-	Multi-Level Inverter
ML Model	-	Machine Learning Model
MLP	-	Multilayer Perceptron Algorithm
MLT	-	Multi-Level Topologies
MARS	-	Multivariate Adaptive Regression Spline
NEA	-	National Environment Agency
NPC	-	Neutral-Point-Clamped
NNARX	-	NN Auto-Regressive Model with Exogenous Inputs
NAM	-	North American -Mesoscale
NWP	-	Numerical Weather Predictions
OD	-	Overall Distribution
PSC	-	Partial Shading Conditions
P&O	-	Perturb & Observe
PLL	-	Phase-Locked Loop
PV	-	Photovoltaic
PI	-	Proportional Integral
RFR	-	Random Forest Regression
RQGPR	-	Rational Quadratic Gaussian Procedure Regression
REST2	-	Reference Evaluation on Solar Transmittance 2

RES	-	Renewable Energy Sources
RMSE	-	Root Mean Square Error
SARIMA	-	Seasonal Auto-Regressive-Integrated-Moving-Average System
SOM	-	Self-Organizing Map
SIMO	-	Single-Input-Multiple-Output
SS	-	Skill Scores
SG	-	Smart Grids
SPP	-	Solar Power Prediction
ST	-	Spatiotemporal
SEVIRI	-	Spinning Enhanced Visible and Infrared Imager
SVC	-	Static VAR Compensators
SVM	-	Support Vector Machine
TDNN	-	Time Delay Neural Network
THD	-	Total Harmonic Distortion
TSI	-	Total Sky Imager
VCO	-	Voltage Controlled Oscillator
VSTSF	-	Very Short-Term Solar Forecasting
WRNN	-	Wavelet RNN

CHAPTER 1

INTRODUCTION

1.1 OVERVIEW

In recent years, traditional energy resources are losing significance due to global warming and the quick exhaustion of fossil fuels. Non-conventional energy resources like solar, wind, biomass, etc, can be highly beneficial in the upcoming days. Incorporating renewable energy sources (RES) with electric grids has gained significant attention and poses several challenging issues among the research community. The discontinuous and uncontrollable nature of solar energy surges the difficulty of grid management and increases the difficulty in handling the production and utilization of electrical energy. Accurate solar power prediction (SPP) models are necessary for optimal grid stability management. The present model mostly concentrates on the solution and the predicting parts that remain unaddressed. The Maximum Power Point Tracking (MPPT) lacks most of the forecasting mechanism; hence it proves to be a generic model. The current MPPT models are inefficient in tracking the maximum power point because of the sunshine variation. The emergence of Artificial Intelligence (AI) technique and their adoption of various techniques have designed more than one domain automatically.

1.2 BASIC INTRODUCTION TO SMART GRIDS

The idea of Smart Grid (SG) combines consumer solutions quantity of technologies and meets numerous regulatory drivers and policies. SG is the next generation power system which utilizes bi-directional flows of information and electricity. The capacity of data amalgamation, reliable data communication, system monitoring, secured data scrutiny, and supervisory and local controls of the SG could fulfil the consumer-supplier demand needs like decline in the power utilization, cost of energy and enhance the efficiency of system. Figure 1.1 illustrates the concept of SGs.

1.2.1 Characteristics of Smart Grids

SG uses innovative services and products added to communication, intellectual control, and monitoring self-healing technologies. The publications recommend the subsequent qualities of the SG.

- SG offers customers better choice of information, and supply further allows customers to act a part in enhancing function of the system. It allows demand response (DR) and demands side management (DSM) via smart gadgets, electricity storage, smart meters, consumer loads, and microgeneration by presenting customers the data about power usage and prices. Incentives and information can be offered to customers to revise their consumption paradigm to overcome some restraints in the energy system and enhance the efficacy.

- It permits function and linking generators of every size and technology and adapts storage gadgets and irregular generation. It assists and accommodates every kind of residential DGs, renewable DERs, and storage choices, micro-generation, thus eliminating the environmental effect of the entire power supply system. It enables 'plug-and-play' function of micro-generators, therefore enhancing the elasticity.

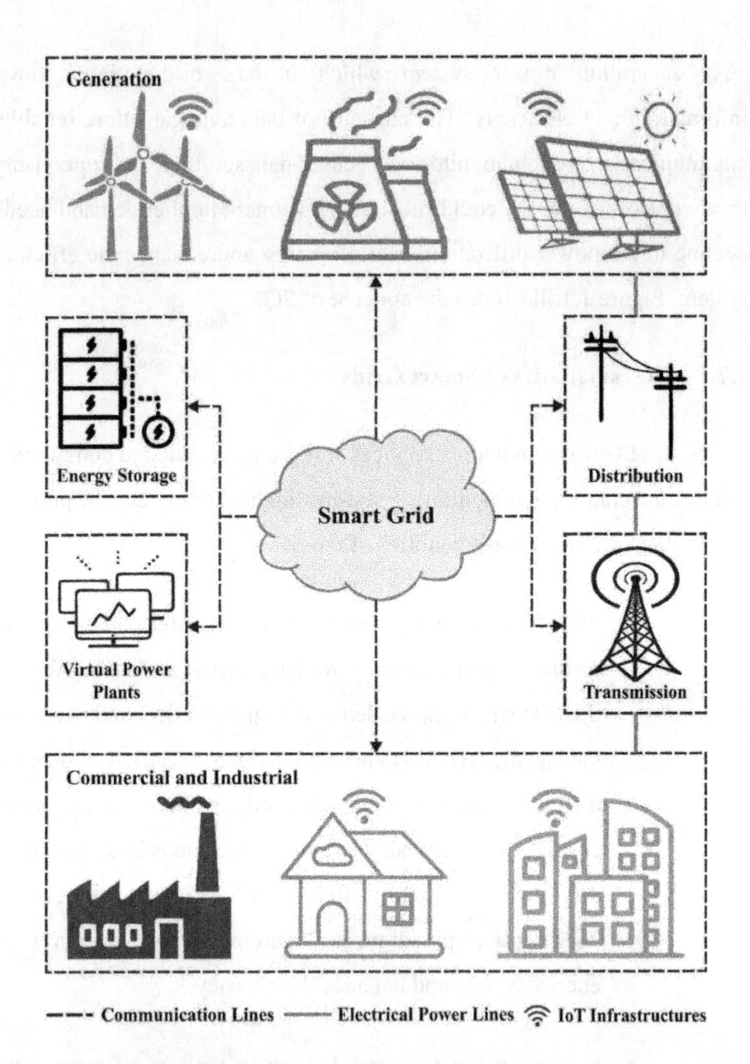

Figure 1.1 Concept of SGs

- It maximizes properties effectively by following effective asset administration and functioning delivery system (operating separately, re-routing power) in relation to the demand. This involves using assets based on what is required and when it is required.

- It works strongly at the time of physical or cyber-assaults, disasters and brings power to customers with advanced stages of reliability and security. It promises and improves reliability and security of supply through reacting and estimating in a self-healing way.

- It offers standards in energy supply to house sensitive apparatus, which advances with the digital economy.

- It initiates to reach the markets via raised DR initiatives, total supply, communication paths, and auxiliary service necessities.

1.2.2 Evolution of Smart Grids

Utility corporations implement identical technology, although they prevail in various changing topographies. Economic, geographical, and Political elements impact the development and erection of the electrical system. Irrespective of these variations, the basic topology of the prevailing electrical power system remains stable. The utility corporations launched numerous command functions and levels of control to alleviate troubleshooting and preservation of the affluent upstream assets. SCADA can be common instance that is broadly used. Nearly ninety percent of every power outage and disturbance contains their roots in the distribution network; from base of the chain, that is, from distribution system, shift against SG must begin. Additionally, the incapability of utilities (utility corporations) to expand their generation capacity in relation to rising electricity demand and bring forth rise in the rate of fossil fuels has fastened the need to update the distribution network by launching innovative technologies that could be helpful with DSM and income protection.

Many recent substructure investments were aimed at metering side of the distribution system. The insertion of the automatic meter reading (AMR) method is considered an instance. AMR in the distribution network permits utilities to learn the status through customers' grounds, alarms, and utilization records distantly. The main disadvantage of AMR system is not able to address DSM. Ability of AMR is limited to reading meter data owing to its single way transmission system. Depending on the data get through the meters, it does not permit utilities in making accurate measures. Conversely, transition to the SG becomes impossible with AMR modules since pervasive control at every level turns out to be impossible with AMR alone. Utility corporations were shifted to AMI instead of capitalizing on AMR. AMI provides utilities with the capability to alter service level variables of customers. Henceforth, growth of electric grid is summarized as: including nerves, adding brains, muscles, and adding bones.

Adding nerves entitles the totalling of sensory gadgets at consumer and utility grid levels. Figure 1.2 showcases the evolution of SGs. The main objective of this is to provide data from the smart choice to the whole system. AMI and Smart meters were customer level nerve systems of SG. At the distribution and communication stage, innovative visualization technologies were used to offer utility grid operators extensive region awareness of grid status.

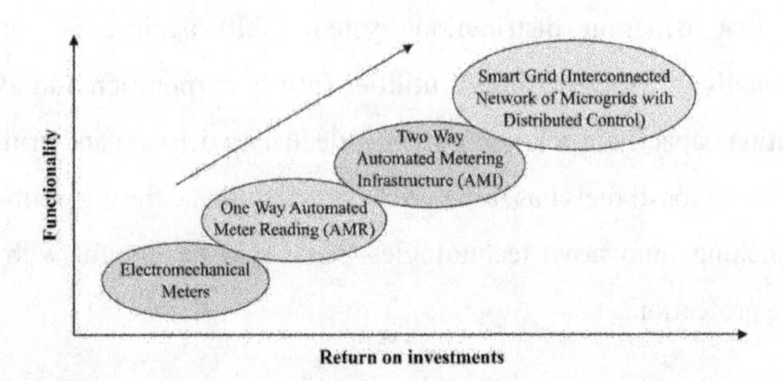

Figure 1.2 Evolution of SGs

1.3 FUNDAMENTAL CONCEPTS OF SOLAR ENERGY SYSTEMS

In this section, some fundamental models of solar power generation and solar irradiation were described, which is simple to understand residual parts of text.

1.3.1 Clear Sky Models

From the abovementioned declaration, the solar irradiance was mostly affected by clouds, whose occurrence complexities irradiance predictive. But it can be feasible to approximate the irradiance in clear sky criteria, i.e., during the nonexistence of clouds. At the same time, the value is utilized for computing the solar indices, normalizing metrics, and attaining the manufacture of solar plants in the stationary states. Generally, the clear sky techniques were provided with weather variables and solar geometry utilizing Radiative Transfer Methods for establishing the links amongst the inputs. There is huge amount of clear sky methods that diverge in everyone, mostly in the input required by all the methods. The most extremely utilized clear sky methods were Reference Evaluation on Solar Transmittance 2 (REST2), Solis, European Solar Radiation Atlas (ESRA), and Ineichen algorithms.

For comprehensive depiction of clear sky methods, reader is represented by the reference as mentioned earlier and to Badescu *et al.* (2013). Several methods only need one input (ESRA, Ineichen), but others need a huge amount (Solis). As comprehensive, the selection of clear sky method to define place was focused on the accessibility and quality of the input dataset, which is an important restraining factor. The chosen of certain methods are of secondary significance. Antonanzas-Torres *et al.* (2016) they developed a calculation of the effect of atmospheric elements from the clear sky approaches at distinct raises and defined its accuracy based on the spatial resolution of inputs.

1.3.2 Clear Sky and Clearness Indices

There exists the clear sky index (k_{cs}) and clearness index (k_t) parameters that are extremely utilized to classify weather situations and calculate smart persistence methods. However, it can be attained in the same manner, diverging from the normalized variable. The clear sky index has been ratio of measuring irradiance for modeling clear sky irradiance I_{cs}.

$$k_{cs} = \frac{I}{I_{cs}} \quad (1.1)$$

The clearness index has been normalizing interms of the extra-terrestrial irradiance I_{ex}. Therefore, it prevents the complexity of modeling the communications of irradiance with atmospheres.

$$k_t = \frac{I}{I_{ex}} \quad (1.2)$$

Whereas $I_{ex} = I_0 \cdot \varepsilon_0 \cdot cos\theta. I_0$, the solar constants get the value of 1360 W/m^2, ε_0 refers to the eccentricity of ellipse explained by Earth from their movement nearby the Sun, and θ indicates the solar zenith angle.

1.3.3 Forecast Definition

Assuming the temporal characteristic of predicts, it can be essential to introduce 3 models, namely forecasting interval f_i, forecasting resolution f_r and forecasting horizon f_h. The forecast horizon is count of time amongst the current time t and effectual time of predictive. The forecast resolution describes the frequency at that the predictive were supplied, and the forecasting interval refers to the predictive time.

1.3.4 Input Origin

Research is classified according to the origin of inputs. Consequently, 2 important techniques are established: techniques that utilize endogenous data created by present and lagged time-sequence of record of the construction of photovoltaic (PV) plant, and techniques which utilize exogenous data that is derived in local measurement (relative humidity, direction, temperature, and wind speed) data in an entire sky imager, satellite image, Numerical Weather Predictions (NWP) (temperature, cloud cover, pressure, irradiance, wind speed and direction, relative humidity, and so on, value in other meteorological database and value in the nearby PV plant.

1.3.5 Persistence Models

The persistence methods are generally utilized as benchmark to further established methods while it can be easiest. It can be considered that criteria (power output, irradiance, clear sky index, and so on) remain the similar amongst t and $t + f_h$. Therefore, some researchers current its outcome utilizing the skill scores (ss) that illustrate the enhancement (or worsening) in terms of the said persistence or baseline systems. But, the comparative interms of skill scores amongst distinct examinations are applied carefully because of the abundance of reference methods. It is only executed if the persistence methods are similar to "Persistence" variables. Then, there exists a list of very general persistence methods. The "naive persistence" considers that the predicted power to time horizon is similar to final value measuring. For sample, for 1hr in advance horizon, the power at 14.00 is similar to the power at 13.00. The difference of this method is for assuming the similar value as to which of preceding day (or neighboring day with accessible measurement) simultaneously. These techniques were recommended if the time series were assumed stationary.

$$P_p(t + f_h) = P(t) \tag{1.3}$$

While the solar irradiance time-sequences weren't stationary, naive persistence was generally restricted to intra-hour application. To overcome this issue, another technique was presented that is executed at extended time horizons, among them in intra-hour application. It has decomposed t solar energy creation from stationary and stochastic elements. The stationary term was commonly linked to clear sky production as well as stochastic term to cloud-induced alters.

$$P(t) = P_{cs}(t) + P_{sf}(t) \tag{1.4}$$

whereas $P_{cs}(t)$ refers to the expected power outcome in clear sky criteria and $P_{st}(t)$ implies the stochastic term. These techniques develop from the supposed "smart persistence," while a further stationary variable was chosen like the ramp persistence, clear sky index, and persistence of cloudiness (Pedro & Coimbra 2012).

$$P_p(t + f_h) = \begin{cases} P_{cs}(t + f_h) & if\ P_{cs}(t) = 0 \\ P_{cs}(t + f_h)\frac{P}{P_{cs}(t)} & otherwise \end{cases} \tag{1.5}$$

In lower variability period and short-time horizon, P_{cs} was highly accurate. Coimbra & Pedro (2013) is also depicted other smart persistence methods dependent upon decomposition mentioned above of solar energy construction, maintenance the stochastic part of time series unvarying.

$$P_p(t + f_h) = P_{cs}(t + f_h) + P_{sf}(t) \tag{1.6}$$

Zhang *et al.* (2015) desired persistence of cloudiness method to its predict.

$$P_p(t + f_h) = P(t) + SPI_t\{P_{cs}(t + f_h) - P_{cs}(f)\} \quad (1.7)$$

whereas SPI_r signifies the solar power index (the ratio of energy to clear sky energy).

In short-time horizon, the ramp persistence has been utilized. It extended the difference from power outcome on the final second for persisting on the predicted horizons.

$$P_p(t + f_h) = P(t) + f_h\{P(t) - P(t - 1s)\} \quad (1.8)$$

1.4 PHOTOVOLTAIC SYSTEMS AND MPPT

A PV process is solid state semiconductor device that generates electrical energy once it is visible to the light. The structure of a solar panel is solar cell. A PV process is produced by interconnecting solar cells in parallel and series. To obtain maximal output voltage, PV module is sequentially interconnected, and to obtain maximal output voltage, they are parallelly interconnected. The solar PV process has been commercialized in several countries as a result of maintenance free and long-term benefits. The main problem in using the PV process is addressing the non-linear features. The PV features depend on the level of temperature and irradiance. PV array experience irradiance levels because of neighbor buildings, trees, or passing clouds. The overall process of PV system has been demonstrated in Figure 1.3.

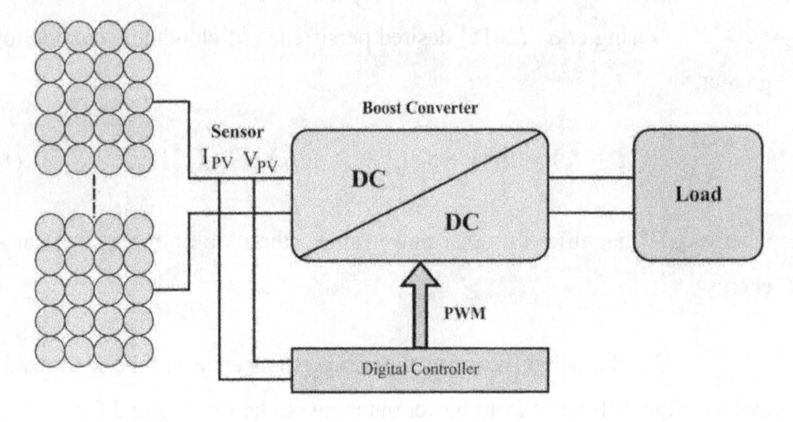

Figure 1.3 Overall process of PV generation system

PV process is categorized according to the component configurations, the equipment interconnected with electric loads, operation, and functionality of systems. They are primarily categorized into standalone, and grid connected systems developed to provide AC and DC electrical supply to function with independence of utility grid that is interconnected with other storage systems and energy resources.

1.4.1 Grid Connected PV System

The grid interconnected PV process is aimed to function parallelly and interconnected with the electrical utility grid systems. They mostly contain an inverter for converting array generated DC into AC with the utility grid's power and voltage quality requirement. A bi-directional interface is made among the PV process AC output circuit and electric utility network at onsite distribution service or panel entrance to permit AC to feed back or supply the grid once the PV scheme output is bigger than load demand. Figure 1.4 demonstrates the overall process of grid connected PV systems. PV scheme can function in standalone, and grid interconnected modes using boost converter and multi-level inverter (MLI) to extract maximal power and feed them to the standalone and utility grid systems.

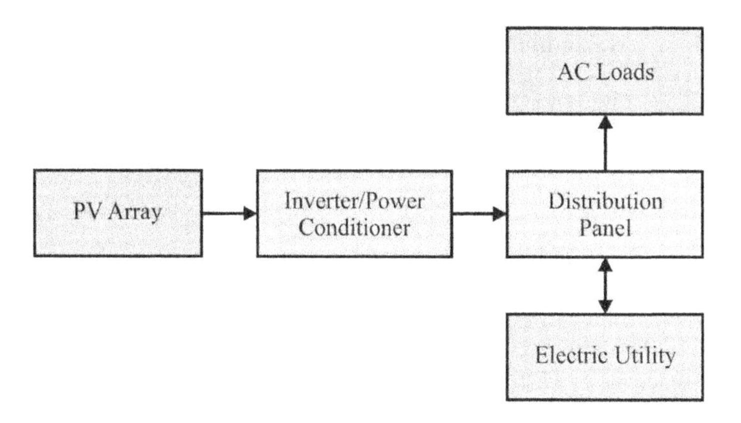

Figure 1.4 Structure of grid connected PV systems

1.4.2 Standalone Photovoltaic System

These are primarily intended to stream certain DC or AC electrical loads and function independently of the electric utility grid. This kind of PV process is known as direct-coupled system. This functions only during sunlight hours, and energy won't be stored in the battery. Water pumps, small pumps, and ventilation fans are the applications for solar thermal water heating systems. Figure 1.5 exhibits the overall process of a direct-coupled PV process.

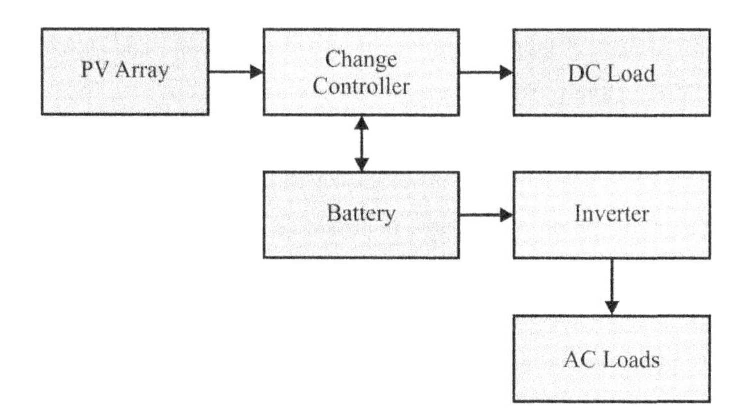

Figure 1.5 Standalone PV systems

1.4.3 Maximum Power Point Tracking (MPPT)

MPPT controlling system is mainly utilized for extracting potential energy of the PV module with corresponding temperature and solar irradiance at certain instant of time by MPPT controlling system. The PV array is unstable, and the voltage and current features curve of PV cell are nonlinear at diverse loads, solar irradiations, and temperatures. The traditional MPPT algorithm is intended for uniform environmental condition, in which the P-V curve generate one MPPT. Bypass diodes, Cloud cover, trees, and buildings cause power peak on the PV string. The major problem of MPPT scheme is to rapidly identify global MPP (GMPP) rather than search for local MPP (LMPP) under partial shading conditions (PSC). MPPT technique examines the efficacy of PV process will be under consideration, and the drawbacks and advantages will be recognized. According to the implementation cost, complexity, flexibility, and reliability, the MPPT method is estimated based on the accuracy and speed of GMPP tracking under PSC.

1.5 SOLAR ENERGY FORECASTING

Conventional energy resource is losing importance because of the rapid depletion of global warming and fossil fuels. Non-conventional energy resources, namely wind, biomass, solar, and so on, are free from the abovementioned challenges. (Peters & Buonassisi, 2019) have analysed the impact of global warming on the efficiency of PV installation. Solar power has become a most promising source to provide energy for industrial, residential, and commercial loads (Ren *et al.* 2015). Nonetheless, the key challenge with RES is unpredictable manner. PV plant generates diverse quantity of energy based on the variations of solar radiation etc. This provides an opportunity to load-generation mismatched in the grids, thus rendering PV energy predicting great importance, especially in grids with higher PV penetration.

1.5.1 Challenges and Need for Solar Power Prediction

Incorporating RES into the electricity grid has greater significance and creates further challenges for researchers and electrical engineers. The uncontrollable and intermittent nature of solar power increases the difficulty of managing grid and balancing the consumption and generation of electric power (Raza *et al.* 2016). A balance between consumption and generation is becoming a serious problem for electrical operators. Low stability, Voltage fluctuation, and lower power quality problems arise due to its uncontrollable nature. Asynchronous, Variability and uncertainty operations are the major problems that electrical operator has to address in incorporating RES with the electric grid (Kroposki 2017). Hence, precise PV energy forecasting is needed for optimum electric grid management.

Solar energy forecasting is needed to reduce the cost of power generation, scheduling, and effective operation of the electric grid, dealing with generated electrical power, congestion management, and approximating the reserves. While the diffusion of solar PV in the grid rises, the forecasting of solar energy becomes increasingly dangerous due to the abovementioned issues. Also, the researcher suggests using a storage system with RE prediction to control power variations. The storage system dampens the fluctuation, maintains a continuous flow of electricity, and absorbs excess power. The primary objective is to compare and discuss forecasting technique for estimating the PV output in two various forms, (i) direct predicting that forecast the energy directly through historical dataset of PV energy and (ii) indirect predicting that make use of solar irradiation forecasting that directly affects solar PV energy generation (Rana *et al.* 2016).

Several approaches for forecasting PV energy were introduced based on physical assumption of the surroundings. NWP is the widely employed physical model of predicting. The unpredictable nature and variation in the atmosphere make the weather prediction method computational complex and severe (Sheng *et al.* 2018). With the development of computer science and its capability to manage nonlinearity, the machine learning technique is gaining popularity. Solar forecasting usually outputs PV power or solar irradiance. The PV generation properties are for solar energy forecasting and modeling. Solar forecasting includes prediction horizon, and related variables are some important characteristics. Standardized performance assessment indices are presented for emerging new solar energy predictors.

1.5.2 PV Generation

The predicted power output of PV generation is influenced by various factors such as estimation of PV cell temperature, the efficacy of inverter measurement of reflectivity, and solar irradiance. The maximal power output is represented as follows

$$P_R = \eta SI[1 - 0.05(t_0 - 25)] \quad (1.9)$$

In Equation (1.9), η indicates the conversion efficacy (%) of solar cell array; S denotes the array area (m^2); I represents the solar radiation (kW/m^2); and t_0 denotes the outside air temperature ©. Usually, MPPT of PV array is a key component of a PV for improving the efficacy. The MPPT approach automatically finds the current I_R and the voltage V_RIn which the PV array efficiently operates to attain the maximal power output P_R in a provided temperature and irradiance, as illustrated in Figure 1.6.

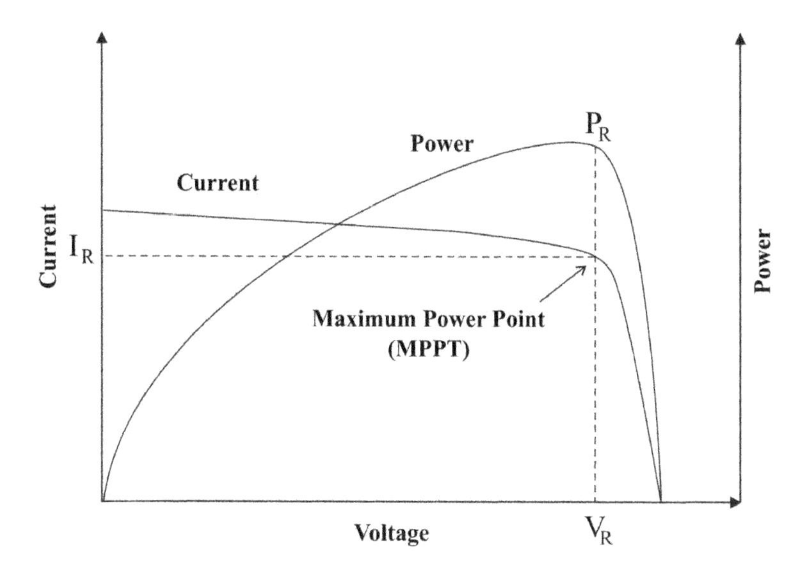

Figure 1.6 Characteristic PV array power curve

1.5.3 Major Aspects of Solar Forecasting

The prediction horizon and election of input variables affect the performance of the predictive method (Lee *et al.* 2018). Generally, the applicable variable available as input of the predictive models of solar power is given in the following:

1) Previous measurement of PV power production;

2) Previous measurement of explanatory parameters like pertinent meteorological variables includes temperature, global horizontal irradiance (GHI), humidity, wind speed, cloud cover, etc.

3) Forecast of explanatory variable, for example, NWP

The input is the available observation of solar energy for forecasting up to 2 *h* ahead, whereas NWP is the input for long horizon.

From the perspective of real-time usage, prediction horizon corresponding to the certain need for decision making activity in the SGs is given by:

1) Very short-term (few seconds to minutes): This forecast is utilized for electricity market clearing, PV, and storage control, namely five minutes for the Australian electrical energy market. In the SG environments, very short-term predicting of solar energy become increasingly significant than before (Dowell & Pinson, 2015).

2) Short-term (up to 4872 hours ahead): This forecast is important for decision-making problems included in the power system operation and electricity market, including unit commitment, economic dispatch, and so on.

3) Medium-term (up to one week in advance): it is helpful for conventional power plants, maintenance scheduling of PV plants, transmission lines, and transformers.

4) Long-term (up to months to years) is utilized for PV plant planning and long-term solar power assessment.

Figure 1.7 demonstrates the Prediction horizon and decision-making activities. From the view of power system operations and SG energy management, short-term and very short-term predictions of solar energy are especially helpful for activities like automatic generation control (AGC), PV plant operations, electricity trading, real-time unit scheduling, and storage control. Many researchers emphasize designing advanced models for short-term and very short-term solar forecasting (Zhu *et al.* 2016).

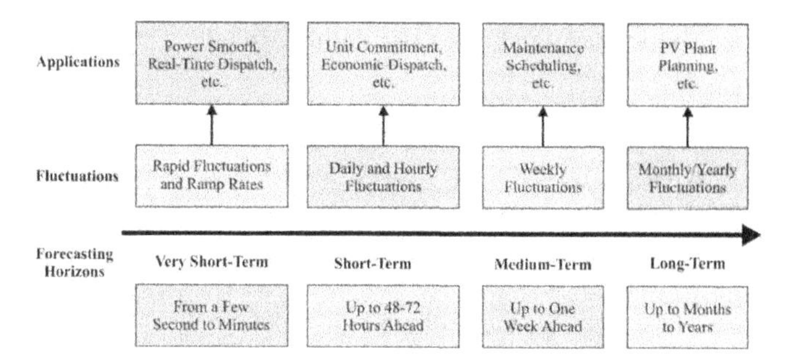

Figure 1.7 Forecasting horizon and decision making activity

1.5.4 Evaluation Measures

Different measurement indices are presented and employed to measure PV and solar predicting performance. Standardized performance measures will be useful for benchmarking and predicting model assessment. The widely employed indices include mean absolute error (MAE), mean bias error (MBE), mean square error (MSE), and root mean square error (RMSE), formulated as,

$$MBE = \frac{1}{N}\sum_{i=1}^{N}[\hat{X}_i - X_i] \tag{1.10}$$

$$MAE = \frac{1}{N}\sum_{i=1}^{N}|\hat{X}_i - X_i| \tag{1.11}$$

$$MSE = \frac{1}{N}\sum_{i=1}^{N}(\hat{X}_i - X_i)^2 \tag{1.12}$$

$$RMSE = \sqrt{\frac{1}{N}\sum_{i=1}^{N}(\hat{X}_i - X_i)^2} \tag{1.13}$$

From the equation, $\hat{X}_i$ and X_i represents the ith observation and prediction values, correspondingly, and N indicates the size of testing data.

The performance measure has its emphasis and own characteristics. The decision maker selects the applicable one for the predictive assessment based on the certain condition.

1.5.5 Types of Solar Power Prediction Models

The present SPP techniques are separated into 4 categories: physical, statistical, regression, and hybrids methodologies, as given in Figure 1.8.

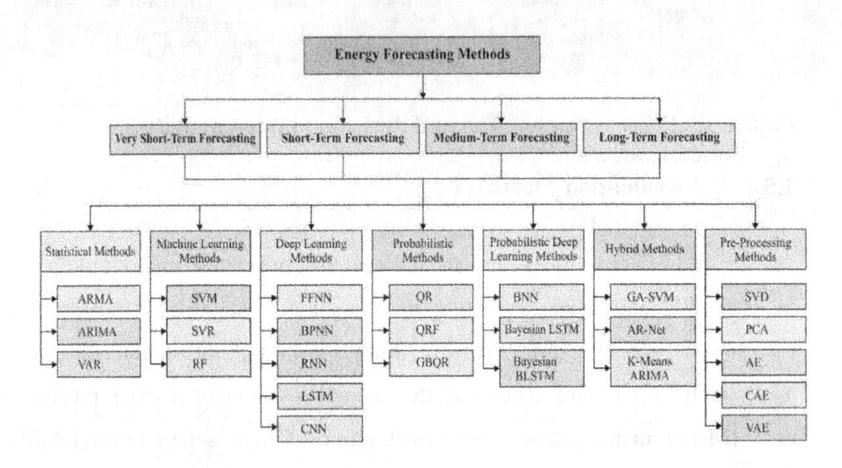

Figure 1.8 Taxonomy of Energy Forecasting Models

1.5.5.1 Physical method

In this kind of investigation, concentrations of SPP are dependent upon NWP and PV cell physical rules (Bakker *et al.* 2019). An input of physical methods has dynamic data like NWP and monitored environmental information and static data like installation angle of PV panels and alteration effectiveness of PV cell. Generally utilized physical methods contain ASHRAE and Hottel. While the physical approaches do not need some historical data, they can depend on geographic data and complete meteorological information of PV panels. Also, the approaches have worse anti-interference ability and could not be dependable for short-term SPP. Figure 1.9 depicts the physical method for PV solar energy forecasting.

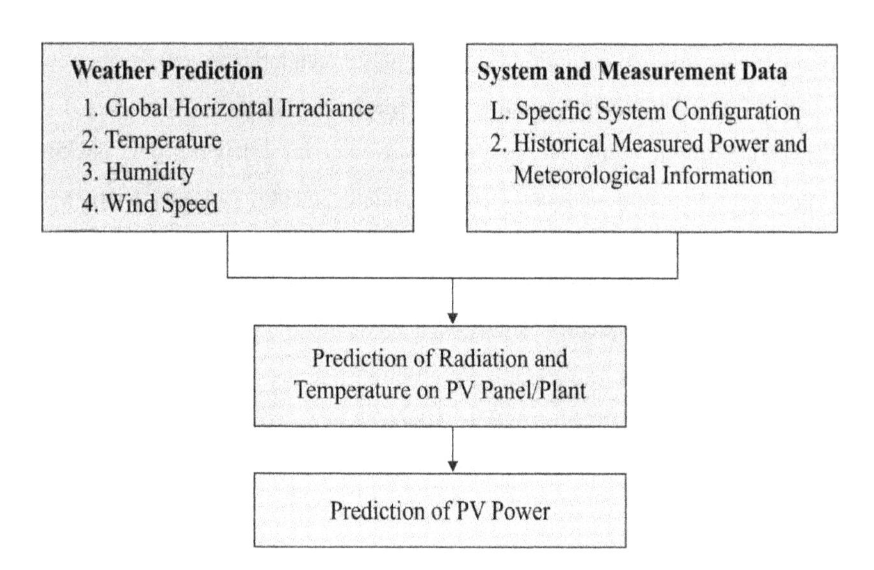

Figure 1.9 Physical approach for PV solar energy forecasting

1.5.5.2 Statistical system

This kind of analysis purposes for establishing mapping connections between historical time-series information and solar power outcome. Generally utilized statistical approaches contain autoregressive (AR) move average, spatiotemporal (ST) association, Kalman filter (KF), grey, and Markov chain model. Usually, the statistical approaches are easy modelling procedures if they are related to physical approaches. A novel statistical approach dependent upon Mycielski and Markov procedures is established for predicting solar irradiance. The probabilistic performance of solar power has been examined by Dong *et al.* (2020), and 2 novel stochastic PV power forecasting methods are presented dependent on stochastic state space method and KF.

1.5.5.3 Regression approach

This kind of examination proceeds with solar radiation, the functioning state of PV panel, and environmental parameters as input variables, and tries to create the scientific connection between input as well as output with curve fits and parameters optimized approaches (Ahmed & Khalid, 2019). The general regression approaches contain wavelet neural network, DL, backward propagation neural network (BPNN), support vector machine (SVM), and extreme learning machine (ELM). A novel forecast framework dependent upon KNN was established for hourly solar radiation forecast. A short-term hybrid forecast method of solar irradiance is dependent upon SVM and Genetic algorithm (GA). A novel forecast structure has been established dependent upon DBN and demonstrated that their forecast efficiency is higher than variation of shallow methods.

1.5.5.4 Hybrid methodology

This kind of investigation concentrates on integrating physical modelling, statistical model, and regression as to hybrid forecast infrastructure. Liu *et al.* (2019) primary embedding PV geographic as well as meteorological information as to spatial mesh, afterward the presented SPP structure dependent upon GRU. A multi-variate hybrid forecast method dependent upon Bayesian average, Elman neural network (ENN) and feedforward neural network (FFNN) are newly presented. The past solar information of University of Queensland has been utilized for verifying the possibility of forecast method.

1.5.6 Applications in Smart Grid Energy Management

With large-scale intrusion of PV, the adverse impacts on the distribution network, particularly on the energy management of SG is grabbing more interest, with difficulties of grid losses, short-circuit current of distributed network, voltage fluctuation, power flow, etc (Kaur *et al.* 2021). PV and Solar energy estimation can offer useful assistance for system operators, electricity participants along with decision makers of electrical power planning. Predicting models having distinct estimation durations is used for SG power supervision. Short-time fluctuation of PV output units becomes very huge, based on the meteorological conditions, like cloud passing. The precise very short-time PV energy estimation method with forecasting time starting from 30 s to more than a few minutes might aid to smooth the PV outputs, ignoring great fluctuations of frequency and voltage of SG. For limiting the ramp rate of PV generations, numerous policies were implied to smooth the PV output unit. Figure 1.10 showcases the applications of solar energy forecasting systems.

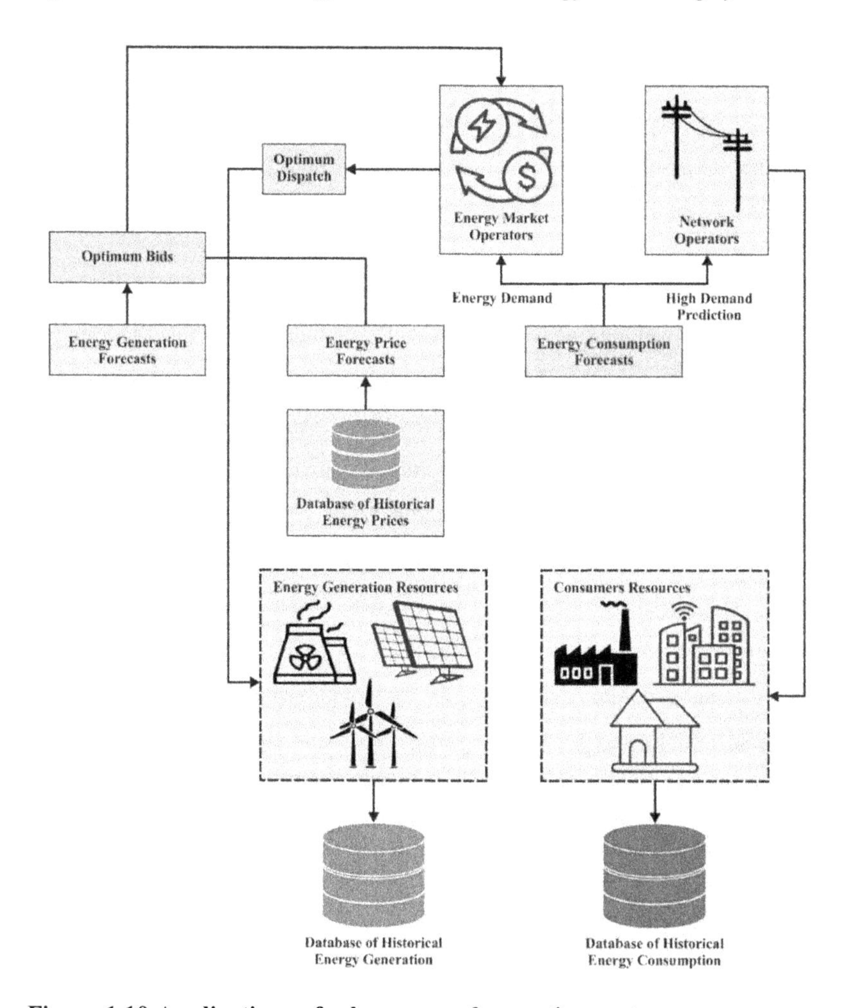

Figure 1.10 Applications of solar energy forecasting systems

An electric double layer capacitor, battery storage sys, electric vehicles, and tem fast ramping generators were typically used technologies for absorption of the fast variations of PV generators (Kaur *et al.* 2018). For programming intraday electric energy of SGs with PV generation integration. If distributed generation (DG) rises in regular procedures, the PV generators would impose important effects on the process of distribution networks, namely distribution network reconfiguration's reliability enhancement, and network loss minimization. Intellectual power management systems have islanded operations and grid-connected that assume the capability and charging rate of storage, distribution network electricity price, and residential load variations. In the SG ambiance, the advancement of day ahead power management tools for next-generation PV settings, involving DR and storage units, leads to uncertainty and flexibility for SG operators (Quan *et al.* 2019).

In hierarchical determinist power management methodology was suggested for fulfilling local power management on the customer side and central energy management of the micro-grid. A price related day-ahead power management structure with DR and storage system for covering the instabilities of the uncertainties of the PV output units. Specifically, the local power management with residential PV process and making-integrated PV microgrid was extensively deliberated by authors. Moreover, day-ahead energy programming turns out to be a significant portion of energy systems assuming the thermal generator's slow ramp limit. The impacts of prediction accuracy of largescale collected PV energy production are analysed (Wang *et al.* 2020). Day-ahead programming of PV production joined with battery storage in the unit commitment issues. One more application of day-ahead estimation method is the bidding strategy of PV corporations taking part in the day ahead electrical energy marketplace.

1.6 MULTI-LEVEL INVERTERS

Nowadays, MLIs are arranged in energy systems due to their capability for meeting the need in power and quality power rating and diminished stage of electromagnetic interference and harmonic distortion. It has numerous benefits of an MLI on the old 2 stage inverters in which high switching frequency PWM was utilized. MLIs were assumed as an industrial resolution for dynamic performance demanding systems and higher-power quality, crossing over a power range of 1-30 MW. Henceforth, MLIs remain models for usage in high power applications as they could produce lower THD output voltage waveforms and could produce high currents having a restricted gadgets grade. The resources of maintainable power, like fuel cells, wind, and PV cells, could widely communicate with a multi-level converter technique. Commonly, the kind of control system used in the PWM of MLIs decides its application, functions, power ratings, and efficiency. Numerous researchers suggested several MLI topologies in the past years (Zhang *et al.* 2017). Still, the typically utilized topologies in the industrial sectors are the cascaded H-bridge (CHB), flying capacitor (FC), and diode clamped or neutral-point-clamped (NPC).

1.6.1 Cascaded H-Bridges MLI

The CHBMLI is generated from the sequential linking of many one-stage H-bridge inverters having individual DC sources (SDCS). Every H-bridge has 4 directional power switches and one DC resource as exposed. Every inverter stage was scheduled for producing 3 voltage output units (−Vdc, +Vdc, and 0,) through the joining of the DC resource to AC output unit; a favourable yield can be attained by linking the 4 switches (S1–S4) in distinct ways. Switching the S4 and S1 switches to the ON position produces the +Vdc output, then if S3 and S2 remain in the ON position, −Vdc output was generated. For producing the 0-output voltage, either S4, S2, and S1 or S3 has to be in ON position.

A serial linking of the AC output units of the complete bridge inverter was done in how the produced voltage waveform would indicate the total of the output units of every inverter. In a cascade inverter, m = 2s + 1 was utilized for representing the amount of output stage power stages, with s as the amount of dissimilar DC resources. This one needs a minor number of elements than FCMLI and DCMLI due to it not having clamping capacitors and clamping diodes. Also, it can be free from the voltage balancing issue due to it won't have DC link capacitors. Conversely, the many DC resources are exchanged by distinct RESs with discrete convertors or through one RESs with multi-output converters in which the voltage balancing can be concerned mostly.

Suggestions were completed for the usage of multi-level cascaded inverters in implementations for static var production, along with that it uses interfaces with RES; it further suggested for utilization in battery-operated empowered applications. A cascade inverter could be utilized for static var compensation through straight linking in sequenced having the electrical system. It is appropriate for the hooking of RES to the AC network as it demands individual DC resources if utilized in PVs and fuel cells. It was suggested to use in electrical automobiles as the chief traction drive due to these implementations numerous batteries assist a SDCSs. The outline of this topologies is elastic, and it is employed in the distinct amount of inverter stages. The production of the distinct output voltages can be achieved through the implementation of distinct ratios of the DC resources and eliminating the inner power levels-based switching idleness. Transformer dependency CHBMLIs were enhanced to eliminate the requirement for autonomous DC resources; it was same as the CHBMLI framework since it varies through the serial joining of the output power of the separation transformer.

1.6.2 Capacitor MLI

Capacitor MLI come closer resemblances within the topology of the DCMLI and FCMLI. The FCMLI depends upon the usage of floating capacitors instead of diodes clamping. For the FCMLI topology, scale of power levels in the output waveform was straight operation of the power disparity happening in the neighbouring condensers. The FCMLI for the 'm' level inverter is made up of 0m to 1 0 DC link capacitors count. Switches S2 and S1 were in the ON position for generating the positive polarity output power whereas S4 and S3 are switched ON for the negative polarity output power. Switches S2 and S4 or S1 and S3 were turned to produce the 0-level output power. In FCMLI, the power synthesis is very elastic than in DCMLI. Whenever there come more than 5 levels, the issue of voltage balancing is met by making a correct range of the switching mixture. One such merit of this topology is activating active power must be managed whereas its disadvantages are in the usage of many capacitors, which made the system costly and hard for assembling. Additionally, the swapping frequency losses in these deployments were more in real energy communication.

1.6.3 Diode Clamped MLI

This inverter termed a neutral point clamped inverter (NPC), is developed by (Nabae *et al.* 1981). The formation of that topology has 4 unidirectional power switches, 2 capacitors, and 2 diodes. The clamping diodes were linked in sequences for sharing the blocking power. In this, the output power is classified at 3 stages, that is Vdc/2, 0, and −Vdc or 2and Vdc or 2 is produced by taking S2 and S1 switches ON whereas S3 and S4 were switched ON for generating −Vdc/2. Switches S2and S3 were turned to yield the zero-level power. At the time of passage of the equal power via the DC link capacitors, it can be anticipated that every dynamic switching gadget contains power stress which can be fastened to the power of every condenser via diode clamping. In practical implementation, the blocking voltage was shared through serially linking the diodes clamping. After which, a voltage stage of V/(m − 1) dc was needed to be impassable by every active gadget. For reversal power blocking, the ratings of power of the diodes clamping should change. If working the DCMLI in the PWM method, the most form of problem that has been modelled in high power implications was the diode reverse retrieval of such diodes clamping.

The DCMLI, compared to other multi-level converter topologies, contains a high industry implementation because of its simplicity, high-power delivery, and capability. It can be discovered in application in high power system interlinks, flexible speed motor drives, and Static VAR Compensators (SVC). But, the issues of this converter involve complexity in one inverter's real power flow because of overcharging or discharging of the DC level without not having enough control, along with the problem of alleviating and corresponding to the capacitor DC power in the DC connection.

1.7 MOTIVATION AND RESEARCH CONTRIBUTIONS

Owing to an increment in the number of consumers and the advent of high-power sectors, electricity production from the energy grid has risen dramatically over the previous few years. Traditional power production has resulted in a large growth in world emissions. As a result, there has been substantial growth in the adoption of solar and wind power in the electricity system. Solar PV systems integrated with a grid in past few years using a phase-locked loop (PLL), string inverter, micro-inverter, etc. Many problems occur which are: a PLL-based clock driver costs two to five times as much as a gate-based clock driver. In PLL, VCO must run due to complex circuits and speeds.

String inverter has single point failure and expandability. If the string inverter fails, then the whole system gets shut down and the rating of the inverter could not be altered once installed. The issue related to solar energy generation is the fluctuations that exist in the produced direct current (DC) because of the displacement of sun and deviance in the number of solar rays from one position to another. This trepidation can be resolved by the use of technical approaches in computing the optimum location of harvesting the solar energy at a faster rate and predicting the quantity of energy production effectively. This research work aims to design PV system with optimal solar power production forecasting and MPPT maximization system. The entire research work is mainly divided into three major objectives as given in the following.

- To develop an AI based on K-Nearest Neighbor (KNN) approach for probabilistic forecasting of solar power production.
- To introduce a new fuzzy based controlling system to control the MPPT method in the solar power generation predicting approach.
- A novel, fickle step-size perturb and observe algorithm (FSPOA) takes place to optimize the maximum MPPT.

1.8 THESIS ORGANIZATION

The rest of the chapters are organized below.

- Chapter 2 shows a detailed review of existing solar power forecasting, MPPT, and MLI techniques available in the literature.
- Chapter 3 focuses on the design of an effective solar power generation prediction model using the AI based KNN model.

- Chapter 4 presents a novel fuzzy based controller to control the MPPT method in the solar power generation forecasting approach.
- Chapter 5 introduces an effective FSPOA technique for the optimization of maximum MPPT.
- Chapter 6 intends to investigate the results obtained by the three different proposed models.
- Chapter 7 derives the key findings of the research work along with possible future enhancements.

CHAPTER 2

LITERATURE REVIEW

2.1 OVERVIEW

This chapter performs a detailed review of existing models related to solar power generation systems. The entire chapter is divided into four major parts. Initially, the traditional solar power generation prediction models are reviewed by including their subclasses. Besides, the recently developed solar power forecasting methods are surveyed briefly. In addition, the recent state of art MPPT models for solar PV systems are elaborated in detail. Finally, the recently presented MLI techniques are surveyed.

2.2 TRADITIONAL SOLAR POWER GENERATION PREDICTION MODELS

In this section, a comprehensive review of traditional approaches to solar power generation is discussed in detail.

2.2.1 Statistical Models

The statistical methods are extremely utilized for time series predicting. Generally, the statistical methods were dependent upon historical data. The predictor focus on creating the connection between the variables utilized as inputs to the statistical method and variable that is forecasted.

2.2.1.1 Persistence

The persistence method was continuously assumed as naive predictor, extremely utilized in meteorology related predicting (Campbell & Diebold, 2005). This easy predicting approach considers that the solar power or irradiance from the future X_{t+1} is the latest measurement X_t formulated as:

$$X_{t+1} = X_t. \tag{2.1}$$

Even if important simplicity, the persistence method was complex that demonstrated for look-ahead time shorter than some hours. The generalizing persistence approach was determined such as the future predicting target is average of final T measure value can be formulated as:

$$X_{t+k} = \frac{1}{T}\sum_{i=0}^{T-1} X_{t-i} \tag{2.2}$$

Whereas is recognized as the moving average (MA). After its simplicity, it is very famous reference method in short-term predicting of wind and solar energy. It can be reasonable that some recently established forecast method is executed superior to some naive reference method; else it could not be meaningful. The persistence prediction accuracy reduces considerably with predicting horizon.

2.2.1.2 ARMA

The auto-regressive moving average (ARMA) is most famous time series predicting method because of its capability for extracting helpful statistical properties (Box *et al.* 2008). In theory, it can be dependent upon 2 elementary parts viz, MA and AR are formulated as:

$$X_t = \sum_{i=1}^{p} \phi_i X_{t-i} + \sum_{i=1}^{q} \theta_i \varepsilon_{t-i} \tag{2.3}$$

whereas X_t implies the predict solar energy or irradiance at time t, p refers to the order of AR method, ϕ_i signifies the ith AR co-efficient, q stands for the order of MA error term, θ_i denotes the jth MA co-efficient and ε define the white noise that is independent variables with constant variance and zero means.

ARMA was generally written as ARMA(p, q), whereas p and q imply the order of AR and MA correspondingly. Mathematically, ARMA (p, q) is changed to $AR(p)$ method if $q = 0$, and an MA (q) method if $p = 0$. The ARMA technique was commonly executed for auto correlated time sequence information and is developed as a common and practical tool to predict the future value of certain time sequences. The ARMA approaches were highly flexible as they can demonstrate many distinct kinds of time series by utilizing distinct orders. It can be demonstrated that fit from predicting if there is a primary linear correlation infrastructure from the time series. The ARMA was executed for predicting future solar generation from California dependent upon the solar radiation data originating in Solar Anywhere and illustrates superior efficiency to the persistence method.

2.2.1.3 ARIMA

An important restriction of ARMA method is that objective time series is stationary, that is, the statistical property of time series does not alter over time. The ARIMA method was established for non-stationary arbitrary procedures. The ARIMA (p, d, q) approach of non-stationary arbitrary procedure X_t is formulated as:

$$\left(1 - \sum_{i=1}^{p} \varphi_i L^i\right)(1 - L)^d X_t = \left(1 + \sum_{i=1}^{q} \theta_i L^i\right)\varepsilon_t \qquad (2.4)$$

whereas L refers to the lag operator determined as $LX_t = Y_{t-1}$, φ_i refers to the AR co-efficient, θ_i denotes the MA co-efficient, ε_t refers to the white noise which is identically distributed and independent arbitrary variables with zero mean, p denotes the order of AR, d implies the amount of non-seasonal variances, and q implies the MA order. During this case of $d = 0$, $ARIMA(p, d, q)$ has changed the ARMA (p, q) method. The ARIMA approach is one of the common class of methods for time sequence predicting. The achievement of ARIMA was due to its exceptional capability for capturing the periodical cycle superior to other approaches. In (Reikard *et al.* 2009), the input data of ARIMA were changed to log values for predicting solar irradiance.

2.2.1.4 ARMAX

In theory, both ARMA and ARIMA could not contain the procedure performance. For considering exogenous input, the autoregressive-moving-average method with exogenous inputs (ARMAX) approach was executed which is demonstrated that a great tool for the time series forecast. ARMAX has been an extension of ARIMA and is further flexible for practical utilization of SPP as it contains external variables like wind speed, temperature, and humidity. This technique is supposed that ARMAX (p, q, b) with p AR terms, q MA terms, and b exogenous inputs terms are demonstrated as:

$$X_t = \sum_{i=1}^{p} \phi_i X_{t-i} + \sum_{i=1}^{q} \theta_i \varepsilon_{t-i} + \sum_{i=1}^{b} \eta_i d_{t-i} \qquad (2.5)$$

In which η_i denotes the parameter of exogenous input d_t.

The ARMAX was presented to PV power predicting and gets to account temperature and humidity as exogenous input which is simply evaluated in the local observatory. It attains superior efficiency to the ARIMA approach (Li *et al.* 2014). A new multi-time scale data-driven prediction method dependent upon spatiotemporal (ST) and autoregressive with exogenous input (ARX) is established to the solar irradiance prediction method. The experimental outcomes that utilize real solar dataset of PV sites from

California and Colorado illustrate the presented method is develop satisfactory outcomes for $1h$ and $2h$ look-ahead times.

2.2.2 Artificial Intelligence Models

AI technique is currently being used in different fields, such as pattern detection, control, optimization, forecasting, etc. With the high regression and learning abilities, AI technique has increasingly been used for prediction and modeling of solar power.

Artificial Neural Network

In theory, multi-layer FFNN could be widespread estimated and have great ability to estimate non-linear mapping to another degree of accuracy. Assume a data with N instances $\{(x_i, t_i)\}_{i=1}^{N}$ whereas activation function ψ, the input $x_i \in R^n$ the output $t_i \in R^m$, and the NN with K hidden node to approximate the N instances are formulated as follows

$$f_K(x_j) = \sum_{i=1}^{K} \beta_i \, \psi(a_i \cdot x_j + b_i), j = 1, \dots, N. \quad (2.6)$$

In Equation (2.6), the weight vector among the ith hidden and input neurons are represented as a_i, the weighted vector among the ith hidden and output units are denoted as β_i, the thresholding value of ith hidden node is represented as b_i, and the output of ith hidden nodes regarding the input x_j is denoted by $\psi(a_i \cdot x_j + b_i)$. The parameter of NN is improved by the distinct models where the BP model is the more commonly used gradient-based model with the objective function is determined as

$$C = \sum_{j=1}^{N} (\sum_{i=1}^{K} \beta_i \, \psi(a_i \cdot x_j + b_i) - t_j)^2. \quad (2.7)$$

In contrast to traditional methods, ANN has been effectively employed in solar prediction. A short-term solar irradiance prediction has been constructed based on BPNN and time series that prevents overfitting and is capable of reaching accurate solar irradiance forecasting Wang *et al.* (2011). The MLP algorithm is employed for predicting the solar irradiance based on twenty-four hours realistic dataset. The presented algorithm provides reference to grid connected photovoltaic (GCPV) and enhances the control algorithm of charge controller.

In Liu *et al.* (2015), a PV power prediction method is presented on the basis of BPNN that considers the aerosol index as a further input parameter to predict the upcoming twenty-four hours of PV power output. The stimulation outcome shows that the betterment of presented method when compared to the conventional ANN method which considers wind speed, temperature, and humidity. In Mellit *et al.* (2014), established three discrete ANN models to apt three standard types of days (overcast, sunny, and partly cloudy) for short-term predicting of the energy produced by a largescale PV plant. ANN model is employed to forecast smaller solar panels for determining the solar prediction horizon for smaller scale solar energy applications. Bayesian neural network (BNN) system is presented to evaluate the day-to-day global solar irradiation with the input variables of relative humidity, air temperature, extra-terrestrial irradiation, and sunshine duration that has greater performance when compared to empirical and traditional NN methods. Wavelet based ANN technique is projected to predict the solar irradiance in Shanghai, which indicates that precise predictions are generated because of the wavelet application (Benghanem & Mellit, 2010).

2.2.3 Other Models

Besides ANN, there are different AI techniques employed to forecast solar power. RBFNN system is applied to the forecasting of the day to day global solar radiation through weather datasets including relative humidity, air temperature, and sunshine duration. A least-square SVM (LS-SVM) based algorithm is projected to predict short-term solar power. The presented method output the predicted atmospheric transmissivity i.e, transformed into solar energy, and outperformed RBFNN based method and a reference AR technique. Various AI approaches like FFNN, linear, RBNNN, ANFIS, and recurrent Elman systems are introduced for the prediction of hourly global solar radiation. The power generation forecasting of PV process is implemented according to the insolation predicting with twenty-four hours advance time through NN, weather information, and fuzzy theory.

A weather-based hybrid model for day-ahead hourly predicting of PV energy system is designed comprising of forecasting, classification, and training phases. Learning vector quantization (LVQ) and Self-organizing map (SOM) are employed for classifying the gathered previous information on PV output. SVR mechanism is utilized for training the output or input dataset of solar irradiance, temperature, and probability of precipitation. Fuzzy inference system is applied for choosing the trained models for precise forecasting. In Moghaddamnia *et al.* (2009), the Gamma test (GT) is integrated into NN auto-regressive model with exogenous inputs (NNARX), local LR, MLP, ANFIS, and ENN to effectively minimize the error and trial workloads. Then, the method is tested on solar radiation. A wavelet RNN (WRNN) is developed for predicting 2-day solar radiation to utilize the relationship among solar radiation and other interrelated variants of temperature, wind speed, and humidity. A hybrid solar radiation forecasting method integrates NNs and fuzzy NN, where temperature information and future sky condition derivative from National

Environment Agency (NEA) are categorized into distinct fuzzy subsets according to the fuzzy rule.

2.2.4 Physical Models

Unlike AI techniques and statistical models, physical model uses PV and solar systems for generating power prediction or solar irradiance.

2.2.4.1 Sky image-based model

The cloud optical depth and cloud cover have crucial impact on solar irradiance. Describing cloud states can be advantageous to solar irradiance prediction. In general, the sky image related model is depending upon analyzing the cloud framework at a specific time. Ground-based sky image approaches and Satellites are utilized for the forecasting of local solar irradiance. Satellites are utilized to forecast local solar irradiance conditions. The satellite image related models is depending on recording and detecting the cloud framework at a specific time and has higher temporal and spatial resolution for predicting solar irradiance. Clouds are discovered and considered from the images for predicting GHI comparatively precisely up to 6 *h* in advance. The time sequence extracted through the analysis of satellite images is utilized for detecting the motion of clouds with the help of motion vector domains.

The short-term predictions of solar irradiance up to 6-h advance have been carried out on the basis of Meteosate satellite images. Resembling predictions can be acquired on the basis of the images of the Geostationary Operational Environment Satellite (Perez *et al.* 2010). A developed method of predicting ground solar irradiance from the satellites (AMESIS) was enhanced with superior accurateness for the incident solar radiation at the surface depending on the spinning enhanced visible and infrared imager (SEVIRI) satellite dimensions. The implementation of innovative sensor nodes like

SEVIRI might enhance the solar estimation accuracy along with the time resolution and high spatial following needs of solar power applications.

On the other hand, the satellite image related methodology, ground related sky images could offer a lot more temporal resolution and higher spatial for solar predictions, based on a total sky imager (TSI). It might recognize the cloud shadow and therefore takes immediate variation in the irradiance, which becomes important for great scale PV power machines or dispersal network confluence having a huge part of PV when one TSI can be used at a site, just short look-ahead time forecasting is attained due to selected spatial scale of cloud pictures and huge cloud changeability. The prediction horizon differs from 5 - to 25 minutes based on the cloud pictures (Chow *et al.* 2011).

2.2.4.2 NWP-related models

NWP is becoming major precise device for solar irradiation prediction with look-ahead time lengthier than numerous hours. NWP system can forecast cloud coverage percentage and solar irradiance based on numerical dynamic modeling of the environment. In theoretical terms, NWP was depending on the accurate information regarding atmosphere state at certain periods and the precise physical laws which govern the atmosphere transition by the fundamental differential formulas.

Generally, NWP offers a lot of advantages when compared to the above-mentioned estimation methods. Satellite imagery systems are appropriate only for 1 - 5 *h* ahead prediction horizon. NWP algorithms were utilized for predicting atmosphere state up to fifteen days in advance. There exists a wide-ranging consensus in which NWPs offer highly precise predictions over satellite-related methodologies in the look-ahead time above 4 hours.

Numerous NMP methods are employed in solar predicting, involving Global Forecast System (GFS), European Centre for Medium Range Weather Forecasts (ECMWF), and North-American -Mesoscale (NAM). ECMWF has been used for predicting regional PV power output in Germany with 3 days look ahead period. In specific, prediction accuracy was raised for regional prediction based on the region size. The ECMWF, NAM, and GFS were authenticated in GHI prediction for the continental United States (CONUS) ground dimension data (Mathiesen & Kleissl, 2011). It was evident that ECMWF contains the highest accurateness in cloudy circumstances, whereas GFS contains best outcome in clear sky situations. A prediction model that depends on grid point value (GPV) dataset was suggested for solar irradiance prediction with the help of three-level cloud covers, relative humidity, and precipitation. Arithmetical research was held in Hitachi and 4 chief cities in Japan, representing the suggested method is dependable. But still, NWP techniques involving ECMWF, NAM, and GFS, consist of certain inherent limits. Owing to the insufficient spatial resolution, it forecast just the average value of grid and does not able to accurately forecast the values of a granted point. Moreover, running NWP needs higher computational costs, i.e, output frequency is 3 *h* for GFS and ECMWF and 1 *h* for NAM because of temporal and spatial limits, the features of many clouds stay as unsolved in NWP with estimation horizons lesser than numerous hours.

2.2.4.3 Hybrid models

Practically, numerous hybrid solar estimation methods were suggested for integrating the advantages of distinct kinds of forecasting methods. A developed model blending nonlinear-autoregressive NN (NARNN) and ARMA a model provides short run predicting global horizontal solar radiation hourly (upto 915 *h* ahead) and predicting high- resolution solar radiation databases (1s - 30*s* scales) with look head time upto 47000 *s* with the help of measured meteorological solar radiation. An original hybrid method including Time Delay Neural Network (TDNN) and ARMA can predict hourly solar radiation offering extraordinary outcomes, in which the ARMA method can be implied for predicting the stationary residual sequences, and TDNN was used for fulfilling the estimation.

The SVM method and seasonal auto-regressive-integrated-moving-average system (SARIMA) were blended for solar power estimation hourly of a small-scale GCPV plant with 20 kWp (Bouzerdoum *et al.* 2013). An original method examining the ARX, clear sky, and AR methods and captivating NWPs as input was advanced for predicting hourly solar power with look ahead time upto 36*h* and verified on 21 PV modules situated in a small town in Denmark. A hybrid prediction method has been suggested by the incorporation of satellite image study for acquiring a cloud cover index through a hybrid exponential-smoothing -state space (ESSS) and method with ANN-SOM. It can be verified on solar irradiance hourly in Singapore, displaying superior outcomes to conventional prediction methods.

2.3 RECENT WORKS ON SOLAR POWER GENERATION PREDICTION

In Belaid & Mellit (2016), a use of SVM for predicting every day and monthly global solar radiation on horizontal surfaces from Ghardaïa (Algeria) was projected. Distinct groups of measuring ambient temperature, computing maximal sunshine period and computed extraterrestrial solar radiation are assumed to be one-step ahead of predicting (one day or month). Baser & Demirhan (2017) concentration on the estimate of yearly mean daily horizontal global solar radiation by utilizing a method that employs fuzzy regression functions with SVM (FRF-SVM). This technique could not be extremely influenced by outlier observation and doesn't undergo over-fitting problems. It can be observed that FRF-SVM method with Gaussian kernel function could not affect by outlier and over-fitting problems and offers the one of the accurate evaluates of horizontal global solar radiation among the executed methods.

In Babar *et al.* (2020), surface solar irradiance evaluates in the ECMWF Reanalysis 5 (ERA5) and Cloud, Albedo, Radiation dataset Edition 2 (CLARA-A2) are utilized as input to random forest regression (RFR) approach for constructing a new dataset with superior precision and accuracy in contrasted with the input dataset. The sky-stratification analysis has been executed and it can be initiated that the presented method offers superior evaluation in every sky criterion with specific enhancements from intermediate-cloudy criteria. In Benali *et al.* (2019), 3 approaches such as smart persistence, ANN, and RF were related to predicting the 3 elements of solar irradiation (diffuse horizontal, global horizontal, and beam normal) measured on the site of Odeillo, France, considered by higher meteorological variables. An objective is for predicting hourly solar irradiation to time horizons in h+1 to h+6.

Bouzgou & Gueymard (2017) presents a novel predicting method for irradiance time series which integrates ELM and mutual information measure. This technique was supposed that Minimum Redundancy – Maximum Relevance (MRMR). For assessing the presented method, their efficiency was estimated against 4 conditions. According to measured irradiance dataset in two dataset sites demonstrating a variety of weathers, the test outcomes expose that chosen of an optimum group of important variables positively affects the predicting efficiency of global solar radiation. Chang & Lu (2018) propose an effectual method for day-ahead predicting PV power output of plant dependent upon DBN integrate with gray theory-based data preprocessor (GTDBN), whereas the DBN tries for learning higher-level abstraction from historical PV outcome data with employing hierarchical manners.

In Cheng (2016), a hybrid solar irradiance now-casting process was presented. The presented hybrid predictor fuses the outcomes in KF and regressor predictors for the benefits of both approaches. The time-varying adaptive system function for KF was planned for managing ramp-down events to predict further accurately. Chen *et al.* (2011) projects the advanced statistical approach for SPP dependent upon AI approaches. The process necessitates input past power dimensions and weather predictions of relative humidity, temperature, and solar irradiance at the site of the PV power system. A SOM has been trained for classifying the local climate type of 24 h in advance given by the online weather services.

Cornejo-Bueno *et al.* (2019) assess the execution of numerous ML in an issue of global solar radiation prediction from geo-stationary satellite data. Distinct kinds of NN, Gaussian Processes, and SVR were chosen as regression methodologies that must be assessed, because of their better outcomes on same issues in the past. For calculating the PV array irradiance and for predicting the PV energy, a physical estimation technique based on solar irradiance on inclined surfaces was suggested Cui *et al.* (2019). Such technique chooses 3 decomposition methods and 4 transposition methods to be integrated into 12 combination prediction methods. Additionally, soiling factor, solar spectral response, and incidence angle was considered in the modified method.

Azimi *et al.* (2016) suggests a hybrid solar irradiance prediction structure with the help of a Transformation related K-means approach, called TB K-means, increasing the prediction accurateness. The presented clustering methodology was a compilation of fresh initialization methods, K-means system, and a new gradual information conversion method. Contrasting to the other K-means related clustering methodologies that are not able to provide a definitive and fixed response because of the selection of distinct cluster centroids for every run, the presented clustering offers continual outcomes for distinct runs of the system.

Aybar-Ruiz *et al.* (2016) provides an original structure for global solar radiation estimation, based on hybrid neural genetic system. In specific an ELM algorithm and grouping genetic algorithm (GGA) were combined in one system, in which the GGA resolves the optimum selection of features, and the ELM performs the estimation. The presented structure was originally due to its usages as input of the system and the output of a numerical climate mesoscale model (WRF) that is atmospheric parameters estimated by the WRF at distinct nodes.

In Jallal *et al.* (2020), an original ML technique called DNN-RODDPSO has been suggested for improving the real-time estimation accurateness of the hourly power produced by 4 dual-axis solar trackers. This method complies with a recent variant of PSO process with new deep neural network (DNN) method termed an arbitrarily arising RODDPSO method. This technique was implemented for search space diversification and for enhancing the training procedure of the DNN method by eliminating the danger of being trapped in local optima.

In Almeida *et al.* (2015), a method utilizing a nonparametric PV method can be suggested, utilizing as inputs numerous predictions of weather parameters from an NWF method, and actual AC power dimensions of PV plants. The method is created on the R atmosphere and utilizes the Quantile Regression Forests as ML tool for forecasting AC power with a confidence interval.

In El-Baz *et al.* (2018), a new methodology for day-ahead PV power generation probabilistic prediction was presented which is particularly optimizing for creating EMS applications. These methods have many elements for developing probabilistic prediction. Primarily, a clear sky system was tuned for incorporating the system and temperature loss and partial shading. The deviation of PV power in the clear sky power was utilized for training a bagging regression tree that generates a deterministic point prediction.

In Durrani *et al.* (2018), accurate and reliable PV power predictions were needed. The PV yield forecast method was projected dependent upon the irradiance predict method and PV system. The PV power prediction was attained in the irradiance prediction utilizing PV method. The presented irradiance predict method was dependent upon several FFNNs. The GHI prediction is a mean absolute percentage error of 23% on a cloudy day and 3.4% on sunny day for Stuttgart.

Eseye *et al.* (2018) presents a hybrid predicting method integrating wavelet transform, PSO, and SVM (Hybrid WT-PSO-SVM) for short-term (one-day-ahead) generation power predicting of real microgrid PV model. The PSO was utilized for optimizing the parameter of SVM for achieving superior predicting accuracy. Fan *et al.* (2020) examined a 3 novel hybrid SVM-PSO, SVM-BAT, and SVM-WOA to predict daily Rd from air-polluted regions. These techniques are more related to standalone SVM, XGBoost, and multivariate adaptive regression spline (MARS) approaches.

In Feng and Zhang (2020), solar predicting namely very short-term solar forecasting (VSTSF) was extremely implemented for assisting power model functions. The VSTSF gets inputs from several sources, among that sky image based VSTSF couldn't yet be well-known related to their counterparts. During this case, a deep CNN (DCNN) approach named SolarNet was established for forecasting the operational 1hr GHI by only utilizing sky images with no numerical measurement and more feature engineering. The SolarNet is a group of methods that create fixed-step GHI from parallel.

Feng *et al.* (2020) estimated a recently established ML approach such as hybrid PSO-ELM, for accurately predicting daily Rs. A recently presented method has been related to 5 other ML approaches such as generalized regression neural networks (GRNN), original ELM, SVM, M5 model tree, and AE in 2 trained conditions. Gao *et al.* (2020) presented a novel CEEMDAN–CNN–LSTM approach for hourly irradiance predicting. Primarily, complete ensemble empirical mode decomposition adaptive noise (CEEMDAN) was utilized for decomposing original historical information as to group of constitutive series for extracting data features. Secondarily, a DL network dependent upon CNN and LSTM was utilized for forecasting solar irradiance from the next hour.

Gigoni *et al.* (2017) widely related to easy predicting methodology with further sophisticated ones on thirty-two PV plants of distinct sizes and technology for an entire year. Besides, it can be attempted for evaluating the influence of weather criteria and weather predictions on the forecast of PV power generation.

Yun *et al.* (2019) presents a prediction network with cascaded network infrastructure, the pre-stage network was utilized for quantifying the weather stimulus features, and post-stage network was utilized for giving the predicting PV power outcome. The presented approach assumes the quantification of weather determining factors, explores the time series intrinsic connects involved from the power output order, and utilizes the unique stacked organization of LSTM infrastructure for reducing the need on scale of trained dataset.

Gupta and Singh (2021) purposes for forecasting the GHI, relating the efficiency of 2 time-series decomposition approaches that is, EEMD and EMD. These approaches divide the original time sequence into group of orthogonal sequences that are named Intrinsic Mode Functions (IMFs).

Additionally, the problem has been changed to supervise learning problems (input-output problems). The feature selection (FS) method chooses the extremely correlated lag variable for reducing the difficulty of method.

Gutierrez-Corea *et al.* (2016) concentrated on utilize of ANNs in short-term forecast of Global Solar Irradiance (GSI). It establishes a novel technique dependent upon observation made in parallel with neighboring sensors and value to distinct variables (humidity, temperature, pressure, wind, and other estimations) utilizing up to 900 inputs (superior dimensional). Khosravi *et al.* (2018) presents the ML approaches for predicting the hourly solar irradiance. The predicting methods are established based on 2 kinds of input data. A primary one utilizes local time, relative humidity, temperature, wind speed, and pressure as input variables of approaches (N1); the secondary one is time-series predicting of solar irradiance (N2) (predicting methods only utilize in past time-sequences solar radiation values for estimating the future value).

Khahro *et al.* (2015) estimate the solar energy resource by introducing diffuse solar radiation approaches and attaining optimal tilt angle to prospective place was southern area of Sindh, Pakistan. Because of the unavailability of measuring diffuse solar radiation data, 9 novel methods dependent upon accessible data in local agency and value attained in present approaches, for predicting diffuse solar radiation on tilted surfaces were introduced.

Jang *et al.* (2016) present an SPP approach dependent upon several satellite images and SVM learning approach. The motion vector of clouds was predicted by employing a satellite image of the atmospheric motion vector (AMV). It can analyze 4 years' historical satellite images and employ them for configuring a huge amount of input as well as output datasets for SVM learning.

Ibrahim and Khatib *2017) examines a novel hybrid system to predict hourly global solar radiation utilizing RFs approach and firefly algorithm (FFA). The hourly meteorological information was utilized for developing the presented approach. The FFA was employed for optimizing the RFs approach by determining an optimum amount of trees and leaves per tree from the forest. Based on the outcomes, the optimum amount of trees and leaves per tree was 493 trees and one leaf per tree from the forests.

Huang *et al.* (2019) present and test a hybrid solar PV power predicting approach that optimum integrates statistical and skycam-based predicts. It can demonstrate the ability of our method for producing accurate predictions seamlessly in 10-s to 10-min ahead utilizing high-frequency measurement from Canberra, Australia.

In Hossain *et al.* (2017), a day ahead and 1hr in advance mean PV output power predicting approach was established dependent upon ELM method. Therefore, the presented predict approach was trained and tested utilizing PO of PV models and another meteorological parameter recording from 3 grid-connected PV approaches mounted on roof-top of PEARL laboratories.

Haque *et al.* (2013) examines a new hybrid intelligent approach for short-term predicting of PV-generated power. This technique utilizes a group of data filter approaches dependent upon WT and a soft computing method dependent upon fuzzy ARTMAP-FA network that is optimization employing an optimized approach dependent upon FFA.

Hassan *et al.* (2017) offers the primary comprehensive analysis for o exploring the potential of tree-based ensemble approaches from modeling solar radiations. The RF, gradient boosting, and bagging approaches are established to evaluate normal radiation, global, and diffuse elements from the daily as well as hourly timescales. The established ensemble approaches are related to their equivalent MLP, SVR, and DT approaches.

2.4 RECENT WORKS ON MPPT TECHNIQUES

Huang *et al.* (2018) presents an altered MPPT technique for PV models in quickly altering PSC. The presented method combines a GA and FFA and enhances their computation procedure using a differential evolution (DE) technique. The GA couldn't be advisable for MPPT due to their difficult computations and minimal accuracy in PSC. During this case, it can be basic the GA computations with combination of DE mutation procedure and FFA attractive procedure. Li *et al.* (2018) presents a new overall distribution (OD) MPPT technique for quickly searching the region neighboring the global MPP that is more combined with PSO MPPT technique for improving the accuracy of MPPT.

Aouchiche *et al.* (2018) examines the group of 2 approaches, the first time is Global MPPT (GMPPT) for 100 kW array. The secondary approach is the Distributed MPPT configuration (DMPPT) for 1 MW PV plant in PSC. This integration purposes for overcoming the disadvantages compared with PSC and to improve the PV model efficiency. A new approach GMPPT controller was presented utilizing Moth-Flame Optimization (MFO) technique as solution to PSC shading.

Tian *et al.* (2020) purposes for analyzing the efficiency of 2 optimized methods. The 2 techniques were PSO and enhanced pigeon technique. This work primary analysis the process of multi-peak outcome features of PV arrays from the difficult environments and afterward presents a multi-peak MPPT technique dependent upon a group of enhanced pigeon population techniques and incremental conductivity approach. An enhanced pigeon technique was utilized for rapidly locating adjacent the MPP, next the step size incremental approach INC (incremental conductance) was utilized for accurately locating the MPP.

In Behera *et al.* (2020), a famous PI controlling system was tuned utilizing spider monkey technique and tested on stand-alone PV model with local load. The presented SMO based PI controlling system improves the P&O MPPT approach for tracking the MPP further quickly and exactly. Mao *et al.* (2018) examines an adapted artificial fish swarm algorithm (AFSA) for MPPT from PV elements in PS. During this approach, the AFSA enhanced by PSO technique with extended memory (PSOEM-FSA) was enhanced with hybridized it with adaptive visual and step, and the resultant technique is a wide-ranging enlargement of AFSA (CIAFSA).

In Abadi *et al.* (2018), the MPPT was effectively planned dependent upon Adaptive Neuro-Fuzzy Inference System (ANFIS) and combined with solar tracking organization for improving the conversion efficacy of PV elements. The planned ANFIS-MPPT approach has of current and voltage sensor, buck-boost converter, and Arduino MEGA 2560 microcontroller as controllers. In Duman *et al.* (2018), a new MPPT approach dependent upon optimizing ANN by utilizing hybrid PSO and gravitational search algorithm based on fuzzy logic (FL) (FPSOGSA) was presented for tracking the function of the PV panel from MPP. The efficiency of presented MPPT algorithm was tested by doing the simulation and experimental analysis in distinct environmental criteria.

Iftikhar *et al.* (2018) presents a non-linear back-stepping controller for harvesting maximum power in PV array utilizing DC-DC buck converting. The regression plane was expressed then gathering the information of PV array in their features curves for providing the reference voltage for MPPT. The asymptotic stability of model was demonstrated utilizing Lyapunov stability condition. The experimental outcomes validate the rapid tracking and effectual efficiency of controllers. Ant colony optimization was tailored for suiting MPPT in PV models and was projected in (Krishnan *et al.* 2020). Artificial ants were utilized from the solution spaces and were developed for foraging and ant that define optimum source of food are retaining but ants fail for searching efficiently were deleted from the populations. The greedy search of potential ants for an optimum food place refers to identifying superior power peaks from the PV model.

Kapić *et al.* (2018) presents a novel approach to MPPT in various environmental criteria like partial shading and sudden alter from irradiance. This technique has 3 parts such as verifying the partial shading rate, searching for global peak (GP) and decreasing the oscillation around MPP. Vicente *et al.* (2020) projects a novel MPPT approach called as irradiance and temperature (I&T) approach. This method utilized the evaluation of irradiance and temperature measurement for defining the MPP. It can be dependent upon the observation of irradiance properties from PV element current and temperature dependency of PV element voltage. The irradiance is estimated with the measurement of short-circuit current.

In Tey *et al.* (2018), an enhanced global searching space DE approach to track the GMPP was established. An important influence of presented method is the subsequent: ability to track GMPP and faster respond against load variation; optimized approach is searching for GMPP in a huge functioning area as it can be executed by utilizing a single-ended primary-inductor converter, and simple tune as minimal parameter is that set from the technique.

Keyrouz (2018) execute Bayesian fusion, an ML approach otherwise utilized for unsupervised classifier, curve recognition, and image segmentation, for achieving global MPPT from recorded time. The experimental outcomes validated with real-life experiment analysis outperformed the ameliorations of presented approach related to recent algorithms. Priyadarshi *et al.* (2019) present, an ANFIS–PSO based hybrid MPPT approach for acquiring quick and higher PV power with zero oscillation track. An inverter control approach was executed through space vector modulation hysteresis present controllers for obtaining quality inverter current with track accurate reference sine-shaped current. The ANFIS-PSO based MPPT approach is no extra sensor condition for measurement of irradiances as well as temperature variables. The utilized approach delivers remarkable drive control for enhancing PV potential removal.

In Dehghani *et al.* (2020), a fuzzy logic controller (FLC) optimizing with a group of PSO and GA was presented for obtaining the MPP. The presented FLC utilizes the ratio of power variation for voltage differences and derivative of power differences for voltage differences as input and utilizes the duty cycle as output. The range of variations from fuzzy membership function and fuzzy rule was presented as an optimized problem optimizing with PSO-GA.

Farayola *et al.* (2018) proposes a new utilization of rational quadratic Gaussian procedure regression (RQGPR) approach for generating the huge and accurate trained dataset needed for MPPT tasks. For confirming the efficiency of RQGPR approach, the group of ANN and RQGPR as ANN-RQGPR methodology outcomes is related to conventional ANN approach outcomes. Macaulay & Zhou (2018) examines a Modified Perturb & Observe (P&O) MPPT technique utilizing FL based variable step size for overcoming any restrictions linked with the P&O MPPT tracking approach for improving the transient response and decreasing the steady-state terminal voltage oscillation. The presented MPPT technique has been executed and testing on indoor emulated PV source which is created in convention solar panel and DC power supply, boost DC-DC converting, and dSPACE-based MPPT controllers.

2.5 RECENTLY DEVELOPED MLT APPROACHES

In Monteiro *et al.* (2018), a new topology of front-end multi-level DC-AC converter was presented for enhancing the combination of renewable energy models as to SGs and preventing power quality problems. The presented conversation was planned for operating as grid-tied inverters, imposing controlled sinusoidal grid power from phase opposition with power grid voltage, and introducing 5 various voltage levels for improving the present waveform. The DC side was appropriate that linked directly to group of PV solar panels with suitable voltage level, or for external DC-DC intermediary converting utilized for interfacing other RES. In Monteiro *et al.* (2019), a new topology of grid-tied converter was presented, considering as important feature the created multi-level voltage (five-levels). The presented grid-tied converting was planned to on-grid interface that is controlled to guarantee sinusoidal presents to every grid voltage criteria. The DC-side is connected directly to DC-DC intermediary converters, responsible for interface RES, as solar PV or wind power system.

Stonier *et al.* (2020) purposes for investigating the removal of harmonics from solar provided cascaded 15 level inverters with support of Proportional Integral (PI), FL, and ANN based controller. Different other approaches, the presented FLC-based method supports attaining decreased harmonic distortion which aims to improvement from power quality. The author presented for providing resultant voltage regulation for maintaining voltage as well as frequency at inverter outcome end from well-suited with grid connection necessities. Suvetha & Seyezhai (2022) concentrations on the unique hybrid power generation, utilizing a new single-switch non-isolated DC–DC converters combined to grid utilizing 9-level inverters. The presented circuit was planned by combining a boost with quadratic boost converter that offers minimal product cost and enhances circuit efficacy. A 9-level inverter utilizing only one input source with lesser amount of switching device was presented with SG application. Jain et al. (2006) presented voltage regulation using STATCOM.

In Pires *et al.* (2018), a multi-level 3 phase voltage source inverter (VSI) to distribute grid-connected PV model was presented. This MLI was dependent upon a novel topology utilizing three 3-phase 2-level VSIs (T3VSI) with isolation transforming. The PV panel is linked at DC side of all the three-phase VSIs. The 3-phase VSIs AC sides were linked to 3-phase isolation transformer with initial open-end winding, making sure multi-level function. In Mukundan *et al.* (2020), a trinary CHBMLI based grid connected solar power transmission model utilizing altered 2nd-order generalization integral control was presented. In 2-stage solar PV, model has single-input-multiple-output (SIMO) single-ended initial inductance converters and 2 CHB infrastructure per phase utilized for verifying the presented method. An altered 2nd-order generalization integral control was projected to active power control and grid synchronization.

Sharma *et al.* (2019) examines an enhanced CHBMLI-based grid-connected hybrid wind-solar energy conversion system (HWSECS) with mandate of power qualities. The wind and solar energy conversion systems were linked separately for isolated DC-link of CHBMLI with its particular DC/DC converter based MPPT method. The presented HWSECS method undergoes the same unbalance voltage as 2 different sources (WECS as well as SECS) were augmentation amongst isolated DC-link.

In Nerubatskyi *et al.* (2020), a comparative analysis of energy parameters of MLIs utilizing several modulation techniques is projected. A necessity of multi-level voltage inverter is for ensuring higher quality output voltage and making sure minimal power loss and maximal performance. A summary of known modulation techniques to output voltage generating from MLI was projected.

Prabaharan & Palanisamy (2016) suggests a single stage MLI outline which conjoins 3 sequences linked full bridge inverter and a one-half bridge inverter for renewable power applications, particularly PV modules. This outline of MLI decreases the value of entire harmonic distortion. The half-bridge inverter used in the presented outline rises the output voltage level to closely double the output voltage level of an orthodox CHBMLI. This high output voltage stage has been produced with smaller amount of power semi-conductor switches than classical configuration, therefore eliminating the entire switching losses and harmonic distortion.

Maheshwari & Chandrasekaran (2019) provides a controlling scheme for reducing the harmonics in a symmetric cascaded MLI- grid interconnected scheme. It was managed by adapted ANN and genetic method. The solar PV acts as input resource and the output voltage is coordinated with grid. The switching angles were obtained so that the important element of output power reserved for steady and lower order harmonics are removed or restricted. The equipped scheme has been synchronized with sun related PV structure to decline the harmonics.

In Motaparthi & Malligunta (2022), an aligned MLI was applied for a PV-Wind related stand-alone system which contains smaller quantity of switches. Both wind speed and solar irradiance were based on climate conditions and are unbalanced, henceforth a battery bank is combined by using a bidirectional DC-DC inverter to offer an uninterrupted energy supply to loads.

In Kenjrawy *et al.* (2022), a nine-level framework of the presented standardized power quality conditioners has been examined. The conditioner is linked among an SG and PV process (UPQC-PV). Followed by, an innovative method for generating switches for the switches of both parallel converters and the series was suggested. This modulation method depends on an adaptive hysteresis band (AHB) which can be decided through an FL controller for acquiring the requisite modified output voltage with minimal distortion.

Bana *et al.* (2020) grants an original declined part count MLI facing single-stage grid-tied PV process and a closed-loop control policy. The suggested MLI contains a level boosting circuit (LBC) and n repeating units which guide in production of 4n + 7 voltage levels in place of 2n + 3 levels. Three distinct methods were suggested for an appropriate choice of DC link powers for enhancing further levels. The application of several DSM methodologies to a variety of loads, including domestic, commercial, and corporate, is examined in Nasir *et al.* (2021) research. The effectiveness of DSM strategies that employ a combination of optimization strategies, such as optimization techniques, heuristic, discounts, and game techniques, is examined. The impact of DSM on systems that include incorporated sustainable power sources, dispersed sources of energy, batteries, and electric cars is also examined.

Panda *et al.* (2021) research is built on current modeling motivated by SGAM, which provides an overall view for merging the models while operating and maintaining security. The presented model supports interoperability challenges, and attainable findings and active research concerns for the SGAM mapping were also highlighted. The concepts provided in this study can be used to build effective and accurate control schemes for SGs that are subject to unknown loading, generation, and transmission restrictions, thereby optimizing and increasing the overall efficiency of the process.

Tchao *et al.* (2021) study outlines critical concerns that must be considered to secure the security of distributed cloud-based power systems. The research also looks at how blockchain and cloud technology could be used to enhance Ghana's limited efficacy transmission & distribution systems. Following that, a design for decentralizing Ghana's grid network is suggested. An interconnected PV-microgrid linked to the state electricity network with the controlled energy plant is used to demonstrate this Bimenyimana *et al.* (2021). Once executed, the suggested technology aims to lower NPC and LCOE, resulting in a cheap and sustainable power system for everyone. The solar power system developed in this work can result in more efficient use of Rwanda's regional energy resources, providing efficient, economical, and ecological access to energy for all people; it is acknowledged as both accurate and economical, as well as having sustainable environmental aspects.

Song *et al.* (2021) research presents a parameter optimization approach for different PV scenarios using an enhanced achievement history adaptation. On computing values of several PV types, the efficiency and optimization techniques technique has been validated. This work is better than the other related algorithms in terms of validity, precision, and computation effectiveness, and the parametric optimization approach is an excellent way for designing, controlling, and upgrading PV systems, according to the empirical and analytical data. As a result, this study is regarded as a valuable tool for improving values in other PV designs.

In Sharew *et al.* (2021) research, the stability of a PV system smart DG based on harmonic distortion is investigated using 34 buses from the distribution feeder. With ETAP technology, the harmonic power movement technique is used to evaluate the voltage range and orders of harmonics. PSO determines that bus 34 is the best site for the solar PV installation. The level of harmonic distortion can be described by the varied loading rates of the solar system coupled with the DG.

An entire electricity production inspection of a small-scale medicine factory is discussed in Lalith *et al.* (2021) study. During analysis, standard drug production technologies complicate the work. As a result, extensive power audition investigations on harmonics are conducted besides altering the production techniques of Indian traditional medicine. The efficiency of the active control is determined using the energy-saving and carbon emission reduction of SMEs after the post-auditing framework and accession while addressing interconnected demands. Furthermore, global climate remediation is guaranteed with the implementation and testing of grid-associated solar systems, which includes energy assessment, cost assessment, and life span carbon pollution of the suggested research. By reducing carbon emissions from SMEs' electricity usage, the suggested solar power plant promotes energy self-sufficiency and renewable power utilization.

In Selvaraj & Victor (2021) research, a sustainable integrated technique is examined and compared to commercial AC refrigeration systems, in which a PV scheme is employed to provide the electric power required to maintain an absorption cycle. The AC and DC systems' Coefficients of Performance (COP) were 0.18 and 0.14, respectively. The system's basic payback period is 10.2 yrs. The development of combined micro water and PV process for remote regions of Indonesia is presented in the (Syahputra & Soesanti, 2020) study. We perform field methods of investigation to identify the perfect quantity of micro water and PV systems, as well as power demand assessment and hybrid power plant optimal design. NASA's dataset for sun energy provided statistics on solar energy capacity. In irrigation canals coming from rivers, water plant data are utilized to create a hybrid power plant that is both efficient and effective.

Electricity generating patterns can be changed to the line more closely with household power usage profiles by installing solar panels at various slopes and angles. The orientation of solar panels is explored in (Laveyne *et al.* 2020) essay to see if it can help with implementation issues in domestic LV distribution systems. The influence of a solar panel installation on a home distribution grid is simulated using an updated simulation tool of a solar panel installation. Real irradiation dataset and actual grid are employed to get as similar to actual conditions as possible.

The purpose of (Mesloub *et al.* 2020) research is to measure the effect of a non-uniform distribution blueprint of DG integrated into a completely glazed open-office facade paired with ILS in cardinal configurations in the climates. To estimate the 3 sections of STPV designs, detailed energy and radiance simulations have been performed. The first group used a-Si particles with various clarities to replace the glass area, whereas the 2nd and 3rd groups used STPVs interconnected with the ILS to replace only 75 percent and 50% of the glazing region, correspondingly.

The solar and the fuel cells are two major RE production resources deliberated in the (Nureddin *et al.* 2020) study. The aforementioned RES-depended energy-producing systems are stationary, power losses are frequently disregarded in comparison to transmission loss in the power grid. For large energy production over changes in the inlet supplies, PV and FC energy systems. An approach for a superior MPPT controlling system is proposed depending on the audited information. This research offers an MPPT technique based on DNNs, which is performed in MATLAB. The major goal of this article was to create the most up-to-date DNN controller for enhancing the power output reliability of a hybrid-PV and FC system. We conducted the assessment in many possible operating situations after creating and testing the suggested system.

The major focus of (Ab-Belkhair *et al.* 2020) article is to offer a new method for PV and wind power generation systems that is dependent on DNN and MPPT and was tested in Matlab software. The microgrid incorporation of a hybrid PV/wind power scheme prerequisite to the generation of DNN controlling system that boosts power quality and decreases THD values. The recommended scheme was evaluated and developed by the Matlab/Simulink tool in different operational circumstances.

Kumar *et al.* (2019) presents a novel SPO methodology for fast MPPT and a novel MMKF approach for PV-grid systems with different loads inclined at the familiar coupling's point for optimal operation. The suggested SPO is a modified version of the P&O technique, which successfully addresses typical PO. As a result, SPO tracks MPP very quickly and collects extreme power from the PV array with great precision. The retrieved power is used to satisfy the active power requirements of loads, and the surplus power is transferred to the grid after the load target is met. At the time of power initializing, the power aspect and power regulation are ensured via the MMKF. However if the grid voltage shows signs of sagging, swelling, THD or dc offset, and so on, the MMKF is utilized in the control scheme to recover basic harmonic components.

For optimal controlling of grid-tied PV systems, Kumar *et al.* (2019) propose a unique LLMLF based modulation technique and LPO-MPPT method. For active component separation from load current, an innovative LLMLF method is designed, as well as a novel LPO MPPT process for optimum MPPT process. The suggested LPO is an upgraded form of the P&O, which effectively mitigates typical P&O computation drawbacks for example steady-state instability, poor dynamic reactions, and limited stepping problems. The primary goal of the designed LLMLF control is to meet the active power requirements of the loads using solar PV electricity generated on-site, with any excess electricity being fed into the grid. When PV power generated is lesser than the essential load power, LLMLF covers the gap by drawing additional power from the grid. The grid's electricity quality improves as a result of this operation. Power factor correction, reactive power compensation, harmonics filtering, and other power quality issues are all addressed by the controller action.

2.6 SUMMARY

This chapter has carried out a comprehensive survey of existing models related to solar power generation systems. The entire chapter is divided into four major parts. In section 2.2, the traditional solar power generation prediction models are reviewed by including their subclasses. Then, in section 2.3, the recently developed solar power forecasting methods are surveyed briefly. Followed by, in section 2.4, the recent state of art MPPT models for solar PV systems are elaborated on in detail. At last, in section 2.5, the recently presented MLI techniques are surveyed.

CHAPTER 3

SOLAR FORECASTING METHODOLOGIES

3.1 OVERVIEW

The conventional ML models for SPP+ face prediction error mainly during the cloudy and rainy days. The state of art ML models are highly generic and integrate distinct model setups. Therefore, this chapter presents probabilistic forecasting of solar power generated by PV cells using KNN algorithm categorized under AI technique. The forecasting process is performed one week ahead regarding current weather conditions and cloud conditions.

3.2 THE PROPOSED SOLAR FORECASTING MODEL

The emergence of Artificial Intelligence (AI) technique and their adoption to variety of techniques has been automatically designed in many different areas. The AI algorithm includes SVM (Laggoun *et al.* 2019) which is supervised mechanism for information processing among the observed and the trained dataset. Excepting SVM, the KNN model (Sun and Huang, 2010) is an effective and supervised mechanism to resolve problematical functions based on the regression and classification function. The KNN model is easily understandable and especially relevant for lesser amounts of data. With the inclusion of noise in the new dataset, the performance of KNN mechanism deteriorates and this may decrease by taking the highest value for "K" which might decrease the noise effect in the new task during the process of classification. This KNN model applied in forecasting solar energy deals with large "K" values.

The most important step in novel KNN algorithm is to define the value of "K", "K" is the number of nearest neighbors ,the optimal value of K is usually found by using the number of samples. The optimal K value is squareroot of N, where N is the number of samples. The AI technique defines the forecasting method of solar energy generated by the PV cells for the next week depending on the dataset gathered from the current week and the information based on the climate conditions. The gathered information is processed by the KNN model of AI technology that is schematically represented in Figure 3.1.

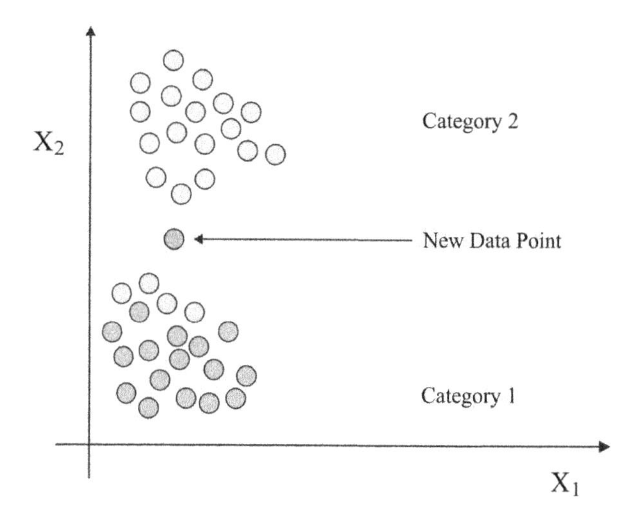

Figure 3.1 K-Nearest Neighbor technique for AI

The figure demonstrates the emergence of 2 dissimilar kinds of collected datasets from the PV cell. The KNN model relates these two datasets to define the novel data point that is emphasized in violet color. The novel data point is derivative get through the 4 neighboring points from class 1 i.e, in green color, and 3 neighboring data points in group 2 that are in yellow. The overall process represents the location of the KNN in the presented method of predicting solar energy produced by the PV cell as portrayed in Figure 3.2.

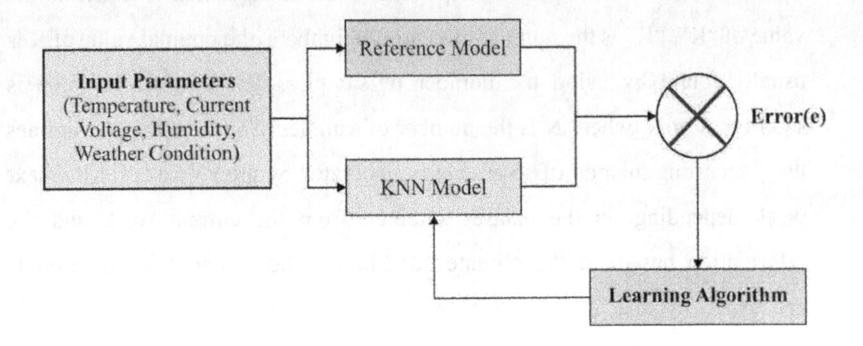

Figure 3.2 KNN Method in solar forecasting process

The figure summaries the forecasting method of solar irradiation using presented AI technique of ANN. The KNN and reference models processed dataset are contrasted to find the generated error in the predicting method is fed back to the KNN algorithm via the Learning model till the error turns into 0. Table 3.1 illustrated the detailed and stepwise processes of KNN model.

Equation (3.1) demonstrates the last predicted value of overall irradiation generated by the PV cell. The value "i" is varied by the amount of solar energy grids existing in the solar power plant. The suggested KNN model should be enhanced and modified to define the solar irradiation by carrying out validation of dataset and weightage of testing dataset. The validity of measured dataset is defined by the following expression.

$$V_d = \frac{1}{K}\sum_{i=1}^{K} S(label\ (x_i), label\ (y_i), label\ (AN_{1,i}), label\ (AN_{1,i}) \qquad (3.1)$$

where the weightage of the testing dataset is implemented by applying the validity of dataset in the measured testing and training dataset. The weighting of testing data is mathematically formulated in the following Equation (3.2).

$$W(x, y) = \frac{x}{d_{E+0.5}} \tag{3.2}$$

Table 3.1 Step -by- Step process of KNN technique in solar irradiation forecasting procedure

Step by Step process of KNN technique in solar irradiation forecasting procedure
Input: Input_Voltage - V_i Input_Current - I_i Temperature - T Humidity - H **Output:** Irradiation (μ) K = 7 **Process:** • Define rules for processing the inputs and determining the nearest relative dataset. 1: If $V_i = A_1$ and $I_i = A_2$; ; $N_i = P_i A_1 + Q_i A_2 + r$ (3.3) 2: If $T_i = B_1$ and $H_i = B_2$; $M_i = P_i B_1 + Q_i B_2 + r$ (3.4) • The data processing can be executed by the KNN framework. Let $A_1 = x_1$ and $B_1 = y_1$; then $f_1 = P_i x_1 + Q_i y_1 + r$ (3.5) Let $A_2 = x_2$ and $B_2 = y_2$; then $f_1 = P_i x_2 + Q_i y_2 + r$ (3.6) • Determine the adaptive nodes with the help of the equations (3.7) and (3.8). The adaptive nodes $AN_{1,i} = \mu_{A_i}(x_i); i = 1,2$ (3.7) The adaptive nodes $AN_{2,i} = \mu_{B_i}(x_i); i = 1,2$ (3.8) • Determining the irradiation (μ_A and μ_B) The irradiation values can be predicted with the help of member functions using Eqs. (3.9) and (3.10).

$$\mu_{A_i}(x_i) = \frac{1}{1+\left[\frac{x-q_i}{o_i^2}\right]^{p_i}} \quad (3.9)$$

$$\mu_{B_i}(x_i) = \frac{1}{1+\left[\frac{x-q_i}{o_i^2}\right]^{p_i}} \quad (3.10)$$

From the expression, o, p, and q represents the precision parameter.
The overall irradiation can be determined with the use of Eq. (3.11).

$$\mu_T = \mu_{A_i}(x_i) + \mu_{B_i}(x_i) = \frac{1}{1+\left[\frac{x-q_i}{O_i^2}\right]^{p_i}} + \frac{1}{1+\left[\frac{x-q_i}{O_i^2}\right]^{p_i}} \quad (3.11)$$

3.3 SUMMARY

In this chapter, a novel KNN based solar radiation prediction is developed and is analyzed quantitatively and is introduced. The prediction task is carried out one week ahead regarding recent cloud and weather conditions. To ensure the goodness of the proposed models, a detailed experimental analysis is carried out using MATLAB 2021a tool. The proposed methodology outperformed well in comparison with the present models and provide higher levels of accuracy with RMSE of 0.576% and consequently provide accuracy of above 99.424% which is regarded to be the more accurate and efficient forecasting methodologies. The effective SPP+ not only enhances the capability of gird connection and security, but it also achieves minimal light discarding.

CHAPTER 4

FUZZY BASED MPPT AND SOLAR FORECASTING USING ARTIFICIAL INTELLIGENCE

4.1 OVERVIEW

This chapter presents an effective hybrid model involving fuzzy based controller for determining MPPT in solar power production. Here, the input signal can be managed by the fuzzification method thereby assigning a value to the input signal. The inference mechanism and fuzzification method produce apparent information according to the set of rules pursued in the inference model. The inference model is sustained through the defuzzification method.

4.2 THE PROPOSED MODEL

The solar energy is produced through metal semiconductor junction PV cell and the MPPT can be accomplished using fuzzy based MPPT controlling system. The MPPT is an electric based DC to DC transformer, applied to boost the matching processes amongst the solar PV cells and the grid utility or the energy saving capacity of the solar energy generating system. The MPPT controlling system is applied during the process of generating PV cell power to transform the higher DC produced from the PV cell to the low DC voltage for storing the energy in the rechargeable battery. Figure 4.1 depicted the graphical depiction of solar PV cell power generation system. The PV cell power generating system using fuzzy based MPPT controlling system permits the process of non-linear controller in the dataset extracted from the proficient database.

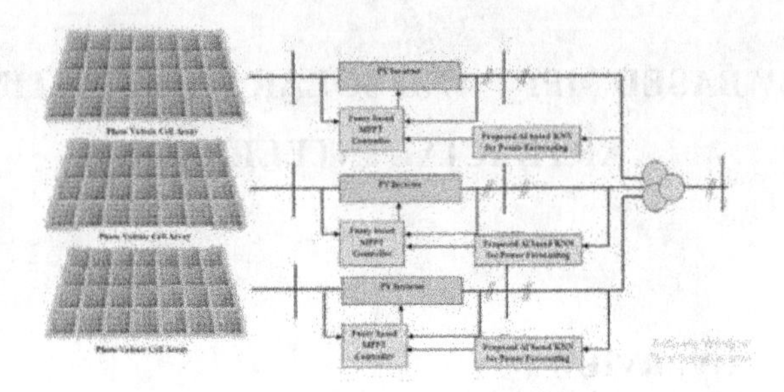

Figure 4.1 PV generation using Fuzzy related MPPT controller

4.2.1 PV Inverter

In this section, the solar energy generating scheme converts the DC produced from the solar PV cell into AC at a rate of consumption. The transformed AC is applied to industrial and commercial applications through electrical grids. The presented PV cell is comprised of the full bridge topology inverter built with 4 switches is applied for synthesizing three level output current waveform and cascaded MLIs that accept current of three dissimilar levels. The mathematical expression that can be attained at the inverter is given in the following.

$$V_o = V_{o.1} + V_{o.2} + V_{o.3} + \cdots + V_{o.n} \tag{4.1}$$

Once the input DC of each PV source is equivalent to the V_{dc}, the inverter is regarded as a symmetrical level. The amount of output current based on the number of bridges built in the inverter circuit is characterized as (2n+1), in which “n” refers to bridge count. The output current of presented model is under control of the fuzzy based controlling system that controls the location of the solar panel for attaining the maximal current from the PV system.

4.2.2 Basics of Fuzzy Logic

FL helps conceptualize the fuzziness in the structure towards a crisp quantifiable variable. Therefore, FL related methods are implemented for efficient power planning to reach realistic solutions. FL handles reality and it can be structured of various valued logic. It addresses reasoning which is estimated with linguistic values instead of crisp values. FL deals with the idea of true value which ranges between totally false and true (0–1). FL was implied in multiple departments. Probability and FL are distinct forms of indicating uncertainty. Fuzzy set theory utilized the idea of fuzzy set membership whereas probability theory utilizes the idea of subjective probability. The many kinds of membership operations generally employed in FL are Γ' function, 'Λ' triangular, Gaussian fuzzy set, '∏' trapezoidal, 'L' function, ''S' function. All such functions are employed in the modelling of power systems. FL related methods in energy systems could array from the simplest to the most complex.

4.2.2.1 Fuzzy models

The FL oriented methods were employed widely owing to their capability to map the realistic condition to a great level. The fuzzy delphi technique has been utilized whenever the response of experts is fuzzy. Numerous rounds are being held amongst specialists to reach a consensus. In fuzzy regression, the information for the independent and dependent parameters was taken in a fuzzy means and the result of the independent parameters on the reliable parameter can be fixed by the derived regression equation. In fuzzy grey prediction, like regression method, the fuzziness can be utilized for capturing the grey region in the parameters assumed for the dependency prediction. Fuzzy ANP and Fuzzy AHP are employed for finding the comparative significance of the parameters.

Fuzzy method is helpful in precisely taking the fuzziness in the people minds when ordering the parameters. In the process of fuzzy clustering, the array provided to the parameters is helpful in obviously defining the clusters and drawing limitations. All these methods were employed based on the problem field. For estimation, fuzzy grey prediction, fuzzy delphi, fuzzy regression, is utilized. Fuzzy ANP, Fuzzy AHP, is employed to find the comparative significance of the power sources. Fuzzy clustering was utilized for clustering sources on the basis of selected criteria namely pollution, cost, availability, and so on. Such methods are categorized as 'simple' as their difficulty is intricated in the method and are utilized for estimation or leveling the significance of power systems.

4.2.2.2 Hybrid models

Neuro-fuzzy techniques and ANFIS techniques have been utilized widely at the time of the past decades, particularly in control systems. The Wide research was carried out in ANFIS and neuro-fuzzy algorithms. Recently, it can be discovered that fuzzy inference modules were employed widely in SG systems or solar PV control. The fuzzy genetic method was also employed in control systems for solar PV or wind and to find the finest wind energy generation terrain. Neuro-fuzzy expert systems and fuzzy expert systems were artificial intelligent systems employed for recognizing the fine energy source or for optimising the existing source. Fuzzy DSS utilizes an integration of several decision techniques. It helps identify the decision method provided an energy state. Fuzzy DEA is considered an optimizing method that helps fix the finest amalgamation of sources which is utilized assuming the several restraints existing in a condition. Such methods can be categorized as 'medium' in complexity. It is employed for precisely simulating the system and its execution. But utilization of such methods is carried out only if the correctness justifies the complexity and cost

4.2.3 Fuzzy based MPPT Controller

The FLC has been explained in detail in Figure 4.2. The input signals are managed by the fuzzification method and allocate values to the input signal. The inference and fuzzification mechanism generate accurate information according to the rules pursued in the inference model. The inference model can be sustained through the procedure of defuzzification.

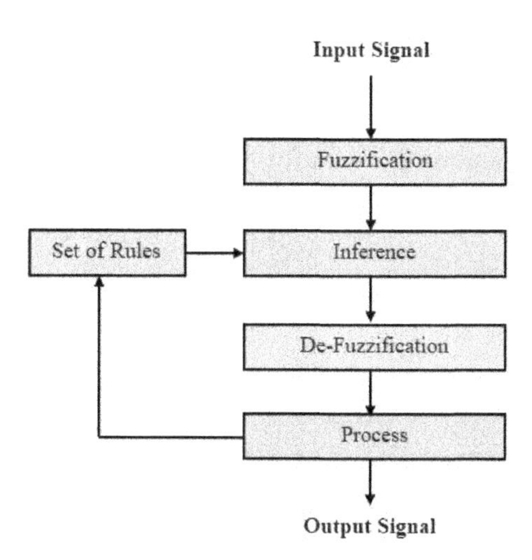

Figure 4.2 Process of Fuzzy related MPPT controller

The procedure implemented in the fuzzy related MPPT controlling system in PV cell power generating system is depending on the de-fuzzification and fuzzification processes are discussed in detail.

4.2.3.1 Fuzzification process

Figure 4.2 depicted the fuzzy controlling system is driven by one process output and two inputs. The two input parameters are referred to as Error and the mathematical expression of Change Error has been given in the following equation.

$$E(t) = \frac{P(t)-P(t-1)}{V(t)-V(t-1)} = \frac{\Delta P}{\Delta V} \quad (4.2)$$

$$E_C = E(t) - E(t-1) = \Delta E \quad (4.3)$$

The input parameter change of Error (E_C) is determined by accomplishing the predetermined displacement condition of the functional point toward the direction of MPPT. The output rises regarding the ΔD changes in duty cycle that value differs from negative to positive array. The output of fuzzy related controlling system was given as an input to DC-to-DC inverter in PV cell generating system that drives the load at constant electrical energy level. The accumulator located at PV is prepared for receiving the ΔD duty cycle value by the following expression.

$$\Delta D = D(t) - D(t-1) \quad (4.4)$$

4.2.3.2 Inference process

The inference mechanism is carried out following the ruleset determined in the fuzzy related MPPT controlling system and Figure 4.3 illustrated the rule table for fuzzy related controller system. The matrix is characterized by the entries based on the value of Error (E), the change in error, and duty ratio value.

Ec / E	**WP**	**RAP**	**PP**	**DAP**
WP	WP	RAP	PP	DAP
RAP	RAP	PP	PP	DAP
PP	PP	PP	PP	PP
WP	WP	RAP	PP	DAP

Figure 4.3 Matrix of Fuzzy Associative Function

Figure 4.3 provides the matrix of fuzzy associative function with forty-nine rule sets for controlling the process of MPPT. The fuzzification method is carried out at four determined scales of power generation system: "PP" Peak Power point, "WP" – Weak Power point and average power point was categorized as "DAP" – Decreasing Average Point and "RAP" – Rising Average Point. The RAP has been plotted if the consecutive power point is in increased way whereas the DAP is plotted when the consecutive power level is in decreased way. The fuzzification method is implemented through comparison made with consecutive power points of 2 solar PV cells. The rule set for plotting the fuzzification table is given below:

WP-WP produces Weak Power point "WP"

WP-RAP produces Rising Average Point "RAP"

WP – PP produces Peak Power point "PP"

WP- DAP produces Decreasing Power Point "DAP"

RAP-RAP produces Peak Power point "PP"

The rule sets are determined from the feedback of output and the columns are referred to as the value of error (E) whereas the rows are referred to as the change of Error (E_C).

4.2.3.3 De-fuzzification process

The de-fuzzification method was regarded as the primary task of fuzzy related MPPT controller system where the output of defuzzification system is given to the PWM system for generating the pulse to drive the semiconductor MOSFET switch in the DC-to-DC inverter. The defuzzification procedure is implemented by two methods such as Centre of Area and Maximal Criterion Method. The Centre of Area methodology defines the controlling output that act as the center of gravity for the previously managed set of fuzzy method. The fuzzy method is defined by carrying out sampling procedure of received dataset and it can be calculated using the following expression.

$$\Delta d = \frac{\sum_{i=0}^{n} \mu(I_i) I_i}{\sum_{j=0}^{n} \mu(I_j)} \tag{4.5}$$

4.2.3.4 Fuzzy based MPPT controller

The presented method drives the DC-DC inverter of the PV cell's power generating process. Figure 4.4 portrayed the proposed technique and it accepts load current, input current, and input voltage to the group of AC into DC converter. The group of ADC admits the analog input and converts them into the corresponding digital value.

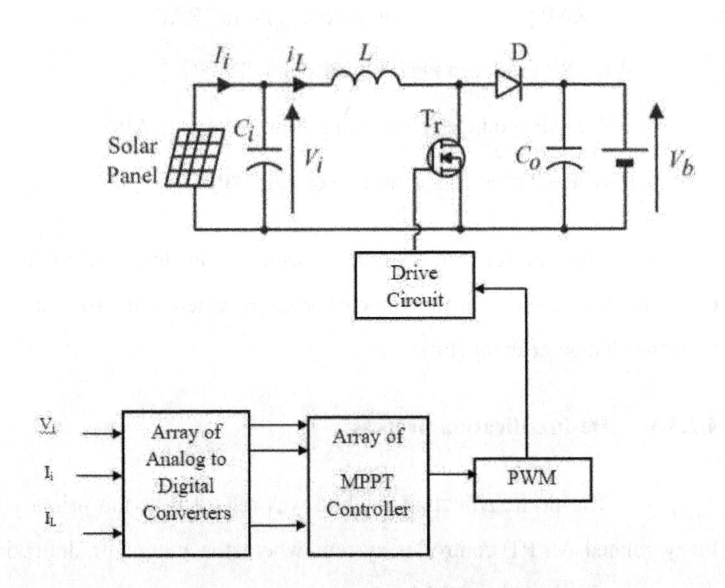

Figure 4.4 Proposed Fuzzy related MPPT controller

The array of MPPT controlling systems encompasses a set of Derivate and MPPT controller systems. The input digital values are adopted by the derived controlling system and the derivative controller frequency is equal to the circuit switching frequency. The MPPT controller undergoes maximal deviation would drive the better Pulse Width Modulation (PWM) resulting in a higher width output pulse prone to the driver circuit. The MOSFET on switching ON, allowing the recent to stream from the source to drain resulting in rotation of solar PV panel. This procedure has been reiterated till the MPPT controller produces a current lesser than the thresholding current thus the PWM output width would be minimal. The mathematical expression of the derived control frequency is shown below.

$$D_c(n) = K_C[D_{iL}(n-1) - D_{iL}(n-2)] \tag{4.6}$$

In Equation (4.6), the derivative coefficient is represented as K_C is and the switching period can be denoted as n. The input current and the input voltage were delivered to the fuzzy related MPPT controlling system for determining the power produced by the PV cell. Figure 4.5 portrays the response of MPPT and the frequency of the MPPT controller is smaller than the LC filter cut-off frequency. The change in switching on frequency $\Delta D_{TON}(n)$ and The derived control frequency $D_C(n)$ control the period of MPPT.

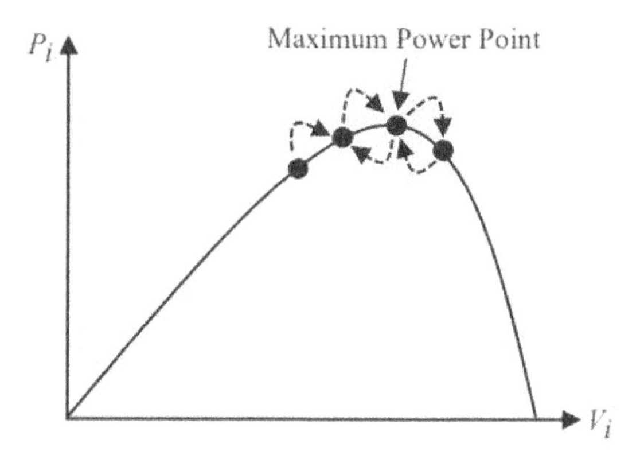

Figure 4.5 Input Power – Voltage Characteristics of Solar PV cell power generation system

4.3 SUMMARY

Existing MPPT models are not sufficient to track the maximal power point because of the variations in sun shine. Therefore, this chapter presents a new fuzzy based MPPT controller and is analyzed quantitatively. The experimental validation of the proposed model is carried out for 6 months with predefined datasets and used IEEE bus systems. In addition, the fuzzy based MPPT controlling system is investigated with current PI controlling system for the critical parameter of THD. The presented fuzzy based MPPT controller obtains effective THD of 1.01% whereas the THD of the PI controller is 3.79%.

CHAPTER 5

INTEGRATION OF PV SYSTEM IN THREE-PHASE SMART GRID USING PWM INVERTER TOPOLOGY

5.1 OVERVIEW

This chapter proposes a novel FSPOA utilized to optimize the maximal power point tracing technique. In the chapter, the PV scheme is integrated with a 3-phase SG using PWM inverter topology. Generally, the conversion efficacy of the PV array is very low. Here, a 3-phase PWM inverter is used for gaining a better output waveform that is slightly similar to the sine wave. We used an auto synchronism tool to match the parameters (voltage, current, frequency, and phase) of the MPPT with these of the 3-phase grid for harmonic elimination to get a better sine waveform. A three-phase LC filter is used for eliminating the unwanted noises in output voltage. For synchronizing the voltage in both of grid and the inverter (i.e, parallel operation), the 3-phase transformer can be utilized. Finally, we get effective PV integration in 3-phase SG with reduced harmonics.

5.2 WORKING PROCESS OF FSPOA MODEL

In this research, the suggested schematic diagram is represented in Figure 5.1. The PV system is integrated with a three-phase SG utilizing a PWM inverter topology in this article. PV array conversion efficiency is often relatively poor. The MPPT converter is a DC-DC inverter that augments the match between the solar array and the energy storage system or power grid.

The FSPOA technique is utilized for optimizes the MPPT output. Here, the PWM inverter is used to achieve a superior output voltage waveform that is comparable to the sine wave. We used an auto synchronism tool to match the parameters (voltage, current, frequency, and phase) of the MPPT with these of the 3-phase grid for harmonic elimination to get a better sine waveform. The 3-phase LC filter is also used to permit desired signals to pass through instead of undesired signals. The 3-phase transformer can be used to synchronize the voltage of both the grid and the inverter.

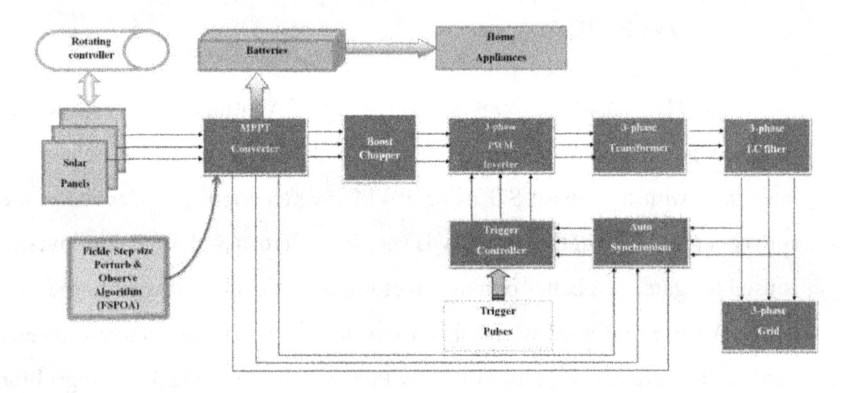

Figure 5.1 Schematic diagram of suggested work

5.2.1 Solar Panels

Solar panels are now utilized in a wide range of electrical devices, such as calculators, that work as long as sunshine is available.

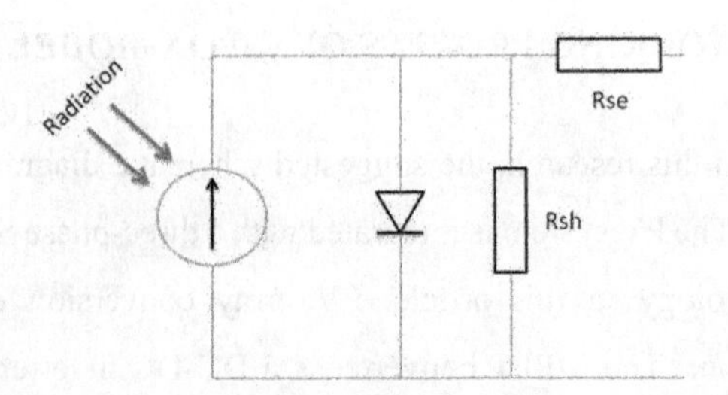

Figure 5.2 Schematic diagram of a cell

Figure 5.2 illustrates the schematic diagram of a cell. The outcome current from the cell is,

$$A_{pv} = A_{ph} - A_D - A_{sh} \tag{5.1}$$

Where,

A_{pv} = PV output current

A_{ph} = photo-generated current

A_D = Diode current

A_{sh} = Current from the shunt connected resistor

$$A_D = A_0 \left\{ e^{c\left(\frac{v_{pv}+A_{pv}R_{se}}{iBt}\right)} - 1 \right\} \tag{5.2}$$

From the Equation (5.2),

A_0 = saturation current

c = charge of an electron

t = temperature

B = Boltzmann's constant

And i = ideality factor of the diode

$$A_{sh} = \frac{v_{pv}+A_{pv}R_{se}}{R_{sh}} \tag{5.3}$$

From the Eq. (5.3),

V_{pv} = outcome voltage of the panel

$$A_{ph} = \left[A_{sh} - t_i\left(t - t_{ref}\right)\right]\frac{L}{L_{ref}} \tag{5.4}$$

From the Equation (5.4),

L = radiation

t_i = shunt connected current component's temperature coefficient

$$A_{pv} = M_p A_{ph} - M_p A_0 \left(e^{c\left(\frac{\frac{v_{pv}}{M_s} + A_{pv}\frac{R_{se}}{M_p}}{iBt}\right)} - 1 \right) - \frac{v_{pv}\frac{M_p}{M_s} + A_{pv}R_{se}}{R_{sh}} \quad (5.5)$$

Where,

Mp = number of shunt-interconnected cells

Ms = number of series-interconnected cells

5.2.2 Rotating Controller

A solar tracking system, for example, rotates the surface of the PV panels or reflecting surfaces to monitor the Sun's motion. Solar trackers can improve solar electrical output by up to 40% when compared to regular panels. Leading to enhanced and much more reliable sun capturing techniques, solar trackers are rapidly being employed in both household and commercial-grade solar panels. Single-axis trackers generally move between east-stream to west-stream, following the Sun's path. The axis of rotation for single-axis trackers is only one angle. This type of tracker has the prospective to boost electricity generation by above 30%. The "main axis" and the "secondary axis" are the two rotating axis degrees on dual-axis trackers.

The revolving axis can shift downwards or upwards during the day to respond to the Sun's angles. Because they can travel in two ways, dual-axis monitors always contact the Sun. Dual-axis tracking is believed to produce 40 percent higher production via energy absorption and provides for the most precise positioning of the solar module. Dual-axis monitors rely on upstream and downstream transitions that are controlled in the same way that solar binoculars are. Dual-axis solar trackers' precise tracking is also utilized in concentrated solar applications like mirrors that direct sunlight catchers and transfer sunlight to energy.

5.2.3 MPPT Converter

MPPT converter is an electronic DC to DC converter that optimizes the comparison between the solar arrays and the battery bank or power grid. Solar cells are fascinating devices. They are, however, not particularly intelligent. Batteries aren't either; in fact, they're downright stupid. The majority of PV panels are designed to produce 12 volts nominally. In reality, nearly all "12-volt" solar panels are meant to generate between 16 and 18 volts. The issue is that a theoretical 12-volt battery is near to a practical 12-volt battery - 10.5 to 12.7 volts, based on charging conditions. Most batteries require between 13.2 and 14.4 volts to completely charge, which is very distinct from what most panels are built for producing. Here, the outcome of the converter provides both the batteries for home appliances and the 3-phase PWM inverter for the mitigation of harmonics. To optimize the converter efficacy, the FSPOA optimization technique was utilized to gain the enlargement of the output.

5.2.4 Fickle step-size Perturb & Observe algorithm (FSPOA)

The constant step size P&O method starts by receiving immediate voltage and current to the outlet panel, then perturbing the operational voltage, calculating the instantaneous power, and comparing it to the prior power. Following that, according to the duty ratio, the voltage level will grow or drop until the maximum power point is attained. If the step size is small, there will be more steady-state errors, and convergence time will be slower, whereas if the step size is large, convergence time will be faster but with more steady-state errors. A different step size P&O approach was designed to handle this issue. Figure 5.3 depicts the FSPOA flow steps.

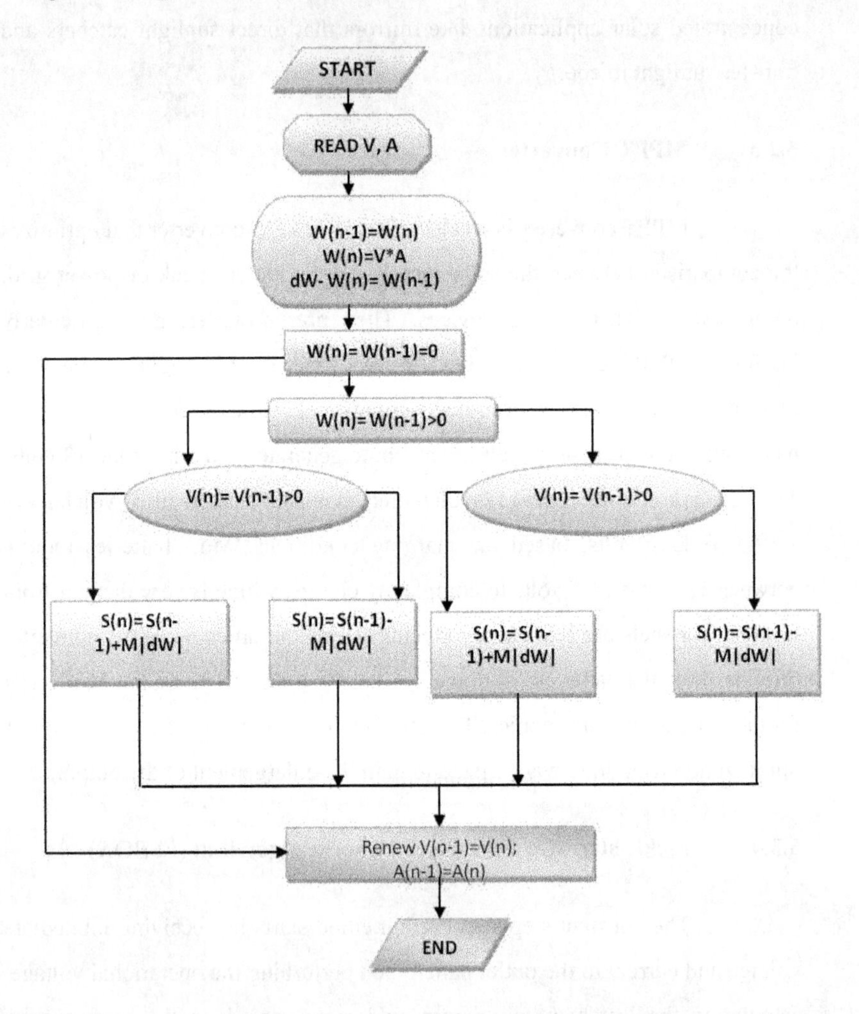

Figure 5.3 FSPOA flow steps

Equation (5.6) defines the step size calculation.

$$S(n) = S(n-1) \pm M \times dW \tag{5.6}$$

From above, S(n) = real value of duty cycle,

S(n-1) = past value of duty cycle,

M = stepping factor and

dW = The power variability concerning modifications in irradiation.

The step size varies naturally in reaction to fickleness in sunlight. Short variations in sunlight result in little fluctuations in the W value and small step size, whereas big fickleness in radiation results in big fickleness in the W value and large step size.

5.2.5 Boost Chopper

Among the most common categories of switching power converters is the boost converter. It receives an input voltage and raises or boosts it, as the term implies. All it comprises is an inductance, a transistor switch, a diode, and a capacitor. Also, a supply of a periodic square wave is mandatory. Figure 5.4 depicts the boost chopper's structure.

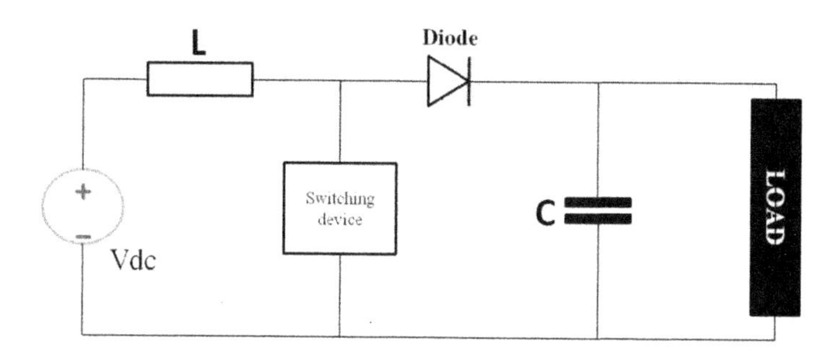

Figure 5.4 Generalized structure of Boost Chopper

The considerable advantages that boost converters provide is their great efficiency - most of them would even reach ninety-nine percent. The incoming energy is turned into valuable electrical output, whereas only 1% is squandered. We have effectively raised the low DC voltage to the desired one because the output capacitor is now charged to a greater voltage than before. To keep the dc output stable under load, these stages are repeated repeatedly. Equation (5.7) defines the average voltage outcome of the boost converter.

$$Average\ output\ voltage = \frac{supply\ voltage}{1-duty\ cycle} \tag{5.7}$$

Where

$$duty\ cycle = \frac{ON\ time}{total\ time} \tag{5.8}$$

5.2.6 3-Phase PWM Inverter

The numbers of three 1-phase 5-level inverters make up the three-phase PWM inverter. The DC line can be utilized to achieve the 5-level voltage. There are fifteen operational controls on the PWM inverter. By joining an extra inverter with an H-bridge inverter, a 1-phase inverter is created. Secondary inverters are made up of {1-IGBT + 4-control diodes}.

The H-bridge inverter in this 3-phase inverter is made up of 4 IGBTs, with Sd3 and Sd4 operating at the reference signal. In the auxiliary circuit, the switches Sd1, Sd2, and Sd5 work at a large frequency. It can be depicted in Figure 5.5. The phase 'd' is a mix of H-bridge inverters (Sd1-S4) and extra inverters (Sd5). Phases 'e', and 'f' were also built in the same way.

Once gate signals from the trigger controller are given, every 1-phase inverter generates 330voltage from the Dc side. The DC line voltage is predictable to be greater when compared to the inverter output voltage (greater than 1.414 times the output voltage). If such a requirement is not met, the inverter will be unfit to send electricity to the load. In the case of a dc connection, the voltage should be larger than fifteen percent. Finally, the 3-

phase step-up transformer receives the outcome voltage of the PWM inverter. The output voltages of a 1-phase inverter are (0, $+V_d$, $+V_d/2$) and (0, $-V_d$, $-V_d/2$).

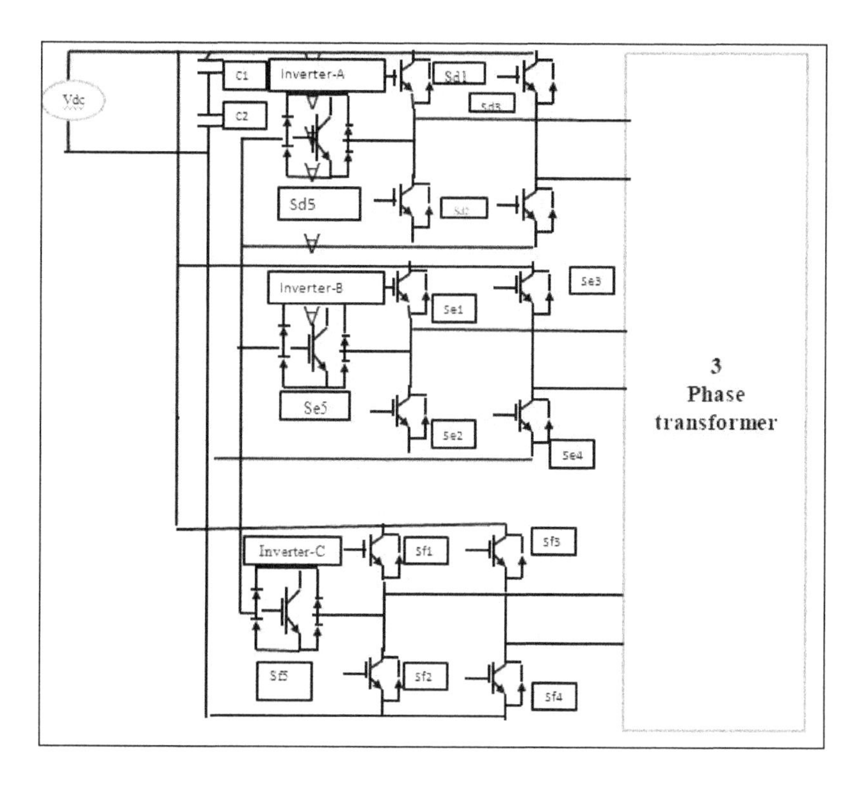

Figure 5.5 3-phase PWM inverter configuration

They may perform in five different modes, each with its voltage level. The valves Sd1, Sd2, and Sd5 have a switching frequency of 20 kHz, while one limb of the H-bridge inverter (Sd3 and Sd4) has a frequency of 50 Hz. The output voltage is identical to a five-level inverter, resulting in a superior harmonic characteristic and fewer passive filters are required. Because each phase is separate, greater results can be obtained. Detecting and replacing damaged devices is fairly simple at the time of fault situation. In this inverter, a random PWM technique is used to mitigate harmonics.

5.2.7 3-Phase Transformer

In this study, the 3-phase transformer is used for voltage synchronization which means the parallel operation of the transformer. In this parallel operation, the inverter output voltage should be equivalent to the 3-phase grid voltage parameter. Like voltage, the frequency and phase are should be the same in both the inverter and the grid. Figure 5.6 depicts the representation of voltage synchronization in both the inverter (V_d, V_e, V_f) and the grid (V_d', V_e', V_f'). Here, the primary and secondary windings of the transformer are connected in star. In this stage, N is denoted as a neutral point.

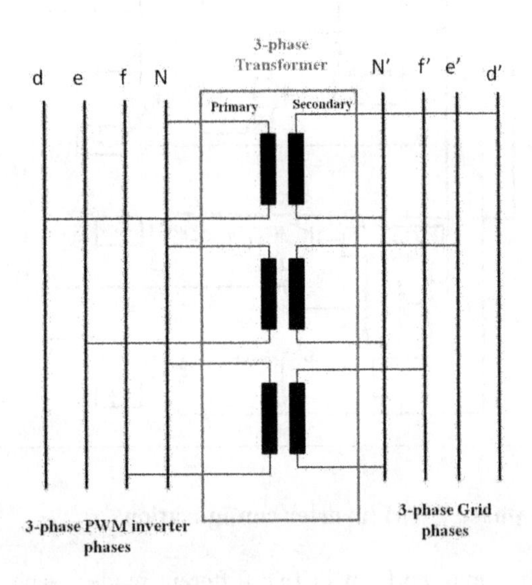

Figure 5.6 Voltage synchronization

5.2.8 3-Phase LC Filter

Normally, the LC filters are mainly used for inverter applications. In this study, the 3-phase LC filter permits only for lower order harmonics in line. Figure 5.7 depicts the 3-phase LC filter connection among both the transformer and the grid.

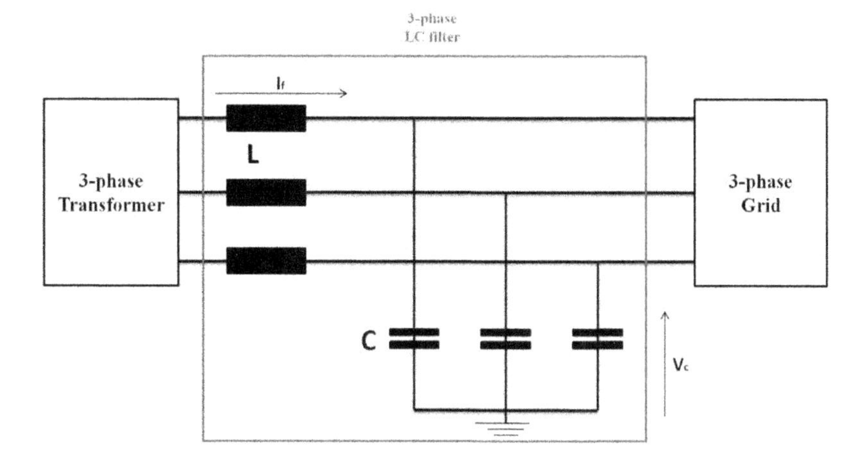

Figure 5.7 LC filter connection

In the above fig, the current flow through the inductor is called I_f and the voltage across the capacitor is denoted as V_c. There are 3 inductors and 3 capacitors for each line included.

5.2.9 Auto Synchronism

In this phase, an auto synchronizer is performed to eliminate the lower order harmonics by synchronizing the parameters of the MPPT with the grid parameters. The comparison result of the synchronizer is then fed to the trigger controller. Whenever the lower order harmonics cross the zero in the waveform, the gate pulse is triggered at that time. By this, we eliminate the lower-order harmonics in the line. Figure 5.8 depicts the algorithm of auto synchronism.

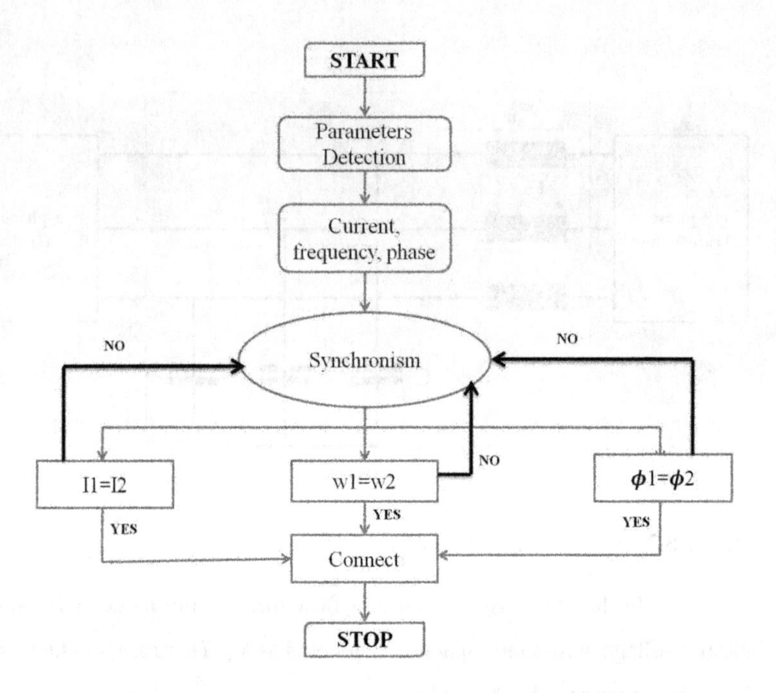

Figure 5.8 Auto-synchronism algorithm

5.2.10 Trigger Controller

Normally, a pulse consists of ON time and OFF time. In ON time, there are several gate-triggered pulses are imported by using the trigger control. The trigger controller produces the trigger pulses concerning the output of the comparator circuit. That means the triggering pulses have been triggered at the same time as the appearance of lower harmonics in the output. Finally, we get the output without lower-order harmonics in PV integration with the grid.

5.3 SUMMARY

In this chapter, a new FSPOA is developed to enhance the optimization of the MPPPT converter. That means different types of step sizes are introduced to optimize the MPPT conversion. To obtain a better sine waveform, the 3-

phase PWM inverter is utilized with a triggering controller. Here also auto synchronism tool was performed to synchronize the parameters of both the MPPT and grid. By this comparison, the trigger controller can trigger the pulses for the elimination of the harmonics. Also, the voltage can be synchronized by utilizing the 3-phase transformer which means the parallel operation was performed.

CHAPTER 6

RESULTS AND DISCUSSION

6.1 OVERVIEW

This chapter focuses on the detailed experimental validation of the three proposed models. The entire chapter is organized into three major parts. Firstly, the suggested model of forecasting the solar irradiation from the solar PV cells using KNN process of AI is designed and executed using MATLAB 2021a. Next, the experimental results for the proposal of fuzzy based MPPT controlling system with the investigation of numerous waveforms are provided in detail. Finally, a detailed analysis of FSPOA model is accomplished using MATLAB tool and the results are extensively investigated.

6.2 RESULT ANALYSIS OF PROPOSED SOLAR POWER FORECASTING MODEL

The quantitative analysis of presented approach of AI based solar irradiation predicting approach was executed on several statistical methods such as MBE, Mean Absolute Percentage Error (MAPE), and RMSE. The listed statistical approaches are arithmetically formulated as in Equations (6.1)-(6.3).

$$MBE = \frac{1}{N}\sum_{i=1}^{N}(\hat{x}(i) - x(i)) \quad (6.1)$$

$$MAPE = \frac{1}{N}\sum_{i=1}^{N}\left|\frac{\hat{x}(i) - x(i)}{x(i)}\right| \quad (6.2)$$

$$RMSE = \sqrt{\frac{1}{N}\sum_{i=1}^{N}(\hat{x}(i) - x(i))} \quad (6.3)$$

Whereas, x(i) refers to the mean value of measured information,

$\hat{x}(i)$ refers to the actual information.

The presented approach of solar irradiation predicting process from solar power generation utilizing PV cells is planned from several approaches regarding 3 IEEE bus methods such as IEEE 57 bus system, IEEE 14 bus system, 23 bus system, and IEEE 30 bus system. The IEEE bus 14 system was regarded as model 1, 23 bus system as model 2, 30 bus system as model 3, and lastly 57 bus system as the model 4.

Table 6.1 Comparison of parameters of multiple IEEE bus systems

KNN Model	MBE	MAPE	RMSE
Model 1	1.1086	1.1156	0.613
Model 2	1.1121	1.1269	0.569
Model 3	1.1978	1.2021	0.591
Model 4	1.2150	1.2021	0.531

To demonstrate the effective solar power forecasting results of the KNN model, a detailed error rate analysis is made in Table 6.1 and Figure 6.1. The results indicated that the proposed KNN model has resulted in a reduced error rate under each model.

Firstly, the error rate analysis of the KNN model is investigated. For MBE, the presented KNN model has provided MBE of 1.1086 on model 1. Similarly, the presented KNN model has attained MBE of 1.1121 on model 2. Moreover, the projected KNN approach has reached MBE of 1.1978 on model

3. Eventually, the proposed KNN algorithm has achieved MBE of 1.2150 on model 4.

To start with, the error rate analysis of the KNN approach was considered. In terms of MAPE, the presented KNN algorithm has provided MAPE of 1.1156 on model 1. Also, the presented KNN approach has attained MAPE of 1.1269 on model 2. Furthermore, the presented KNN model has obtained MAPE of 1.2021 on model 3. Finally, the presented KNN algorithm has attained MAPE of 1.2021 on model 4. To begin with, the error rate analysis of the KNN model is investigated. In terms of RMSE, the proposed KNN approach has offered RMSE of 0.613 on model 1. Similarly, the proposed KNN system has attained RMSE of 0.569 on model 2. Moreover, the presented KNN method has obtained RMSE of 0.591 on model 3. At last, the presented KNN methodology has gained RMSE of 0.531 on model 4.

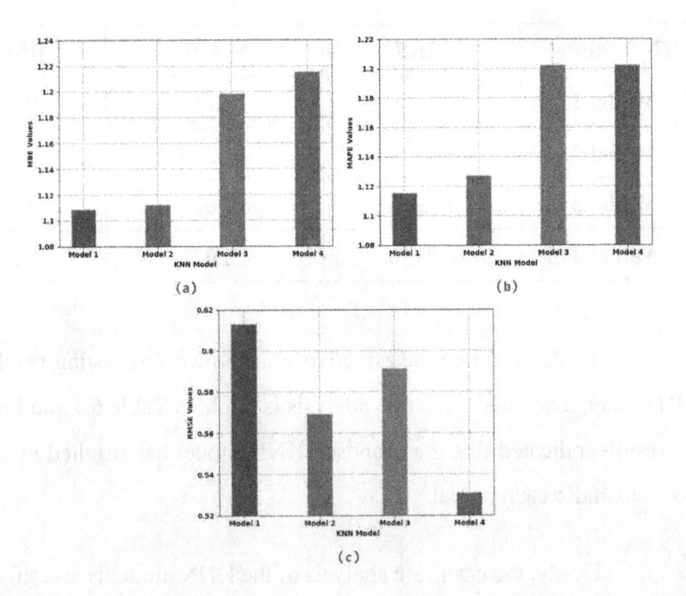

Figure 6.1 Comparison of Error for multiple IEEE bus models using KNN algorithm

Table 6.2 and Figure 6.2 highlight the comparative RMSE analysis of the presented AI based KNN model with other existing models such as ML and DL models. The experimental outcomes stated that the presented AI based KNN model has resulted in superior results with minimal values of RMSE under all IEEE bus models.

In model 1, the presented AI based KNN model has attained lower RMSE of 0.613. At the same time, the existing DL model has offered slightly increased RMSE of 0.627. Followed by, the ML model has resulted in higher RMSE of 0.653.

Table 6.2 Comparison of RMSE with present methodologies

IEEE Bus Model	ML	DL	Proposed AI (KNN)
Model 1	0.653	0.627	0.613
Model 2	0.619	0.599	0.569
Model 3	0.603	0.599	0.591
Model 4	0.579	0.561	0.531

In model 2, the presented AI based KNN algorithm has reached minimal RMSE of 0.569. Also, the existing DL approach has presented somewhat higher RMSE of 0.599. Next, the ML algorithm has resulted in superior RMSE of 0.619. In model 3, the projected AI based KNN algorithm has attained lesser RMSE of 0.591. Besides, the existing DL model has presented somewhat increased RMSE of 0.599. Followed by, the ML model has resulted in higher RMSE of 0.603. In model 4, the presented AI based KNN methodology has attained lower RMSE of 0.531. Moreover, the existing DL model has offered slightly higher RMSE of 0.561. Followed this, the ML algorithm has resulted in superior RMSE of 0.579.

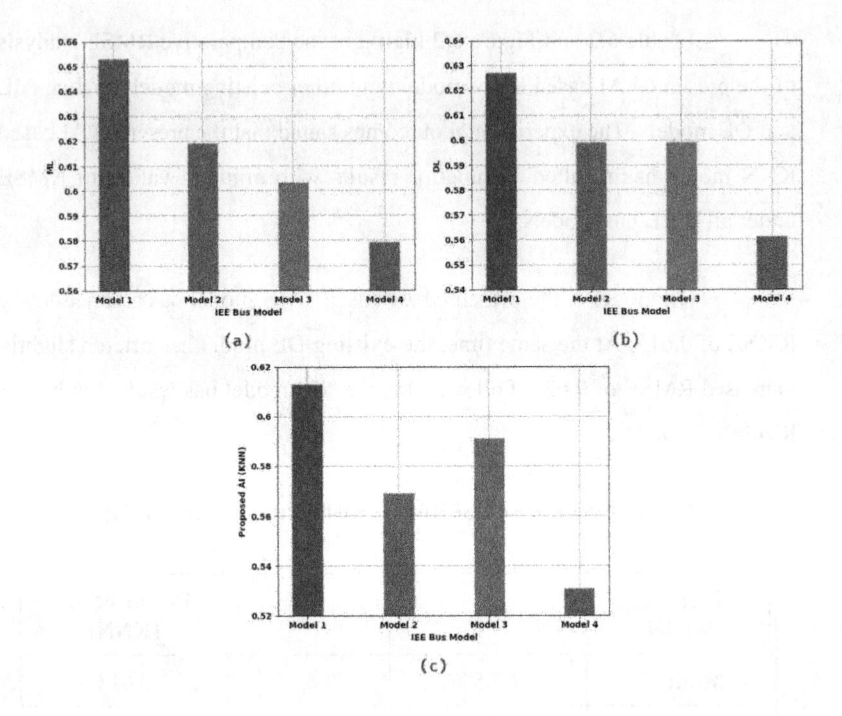

Figure 6.2 Comparison of RMSE value with existing forecasting methodologies

From these results, it is apparent that the presented AI based KNN model has resulted in enhanced performance over other solar energy generation prediction models.

6.3 RESULT ANALYSIS OF FUZZY BASED MPPT CONTROLLER FOR PV SYSTEM

The suggested approach of fuzzy based MPPT control system is developed and performed by MATLAB 2021a. The resulting segment is two-folded with the numerous waveforms in initial fold and afterward the quantitative assessment of the presented methodology in the second fold.

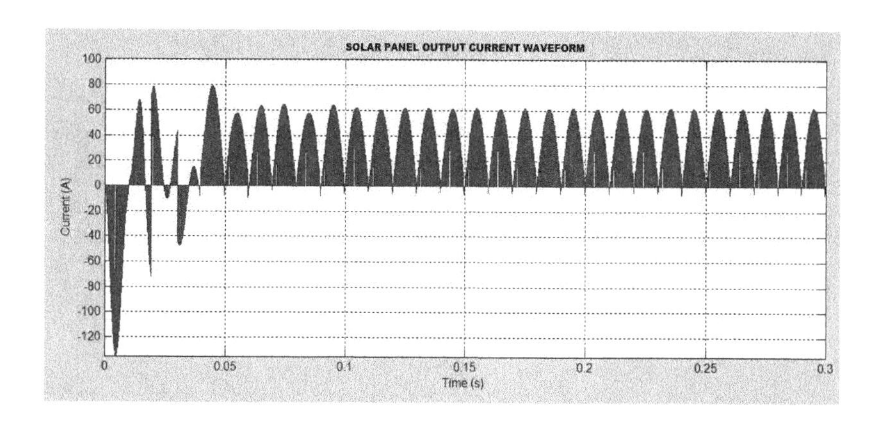

Figure 6.3 Solar Panel Output Current Waveform

Figure 6.3 portrays the output voltage produced by the PV cell. The output current and voltage produced by the PV cell is depending on the amount of irradiation incidents on the PV cell. The voltage generated by the solar panel has fluctuation and produces a constant output current.

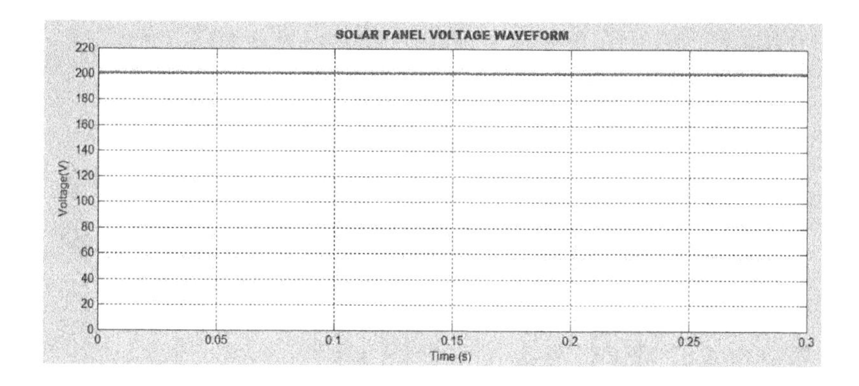

Figure 6.4 Solar Panel Output Voltage Waveform

The overall grid voltage and current produced by the solar grid are depicted in Figure 6.4, 6.5, 6.6, and 6.7 which illustrate constant fluctuation.

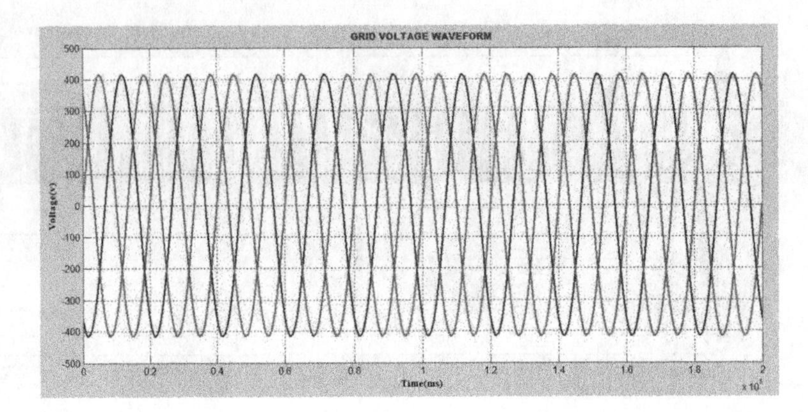

Figure 6.5 Grid Voltage Waveform

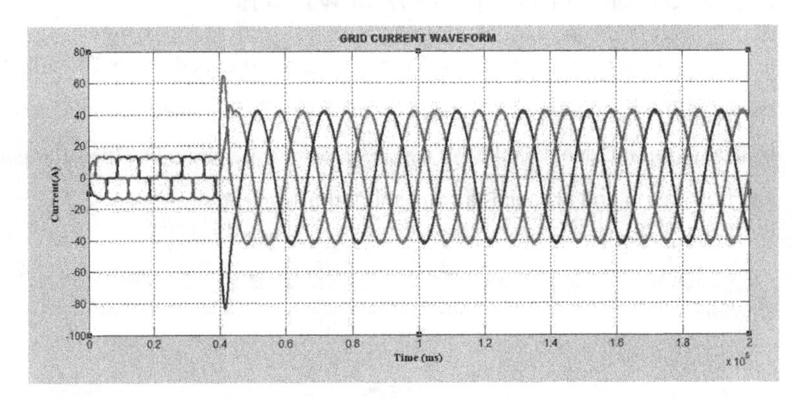

Figure 6.6 Grid Current Waveform

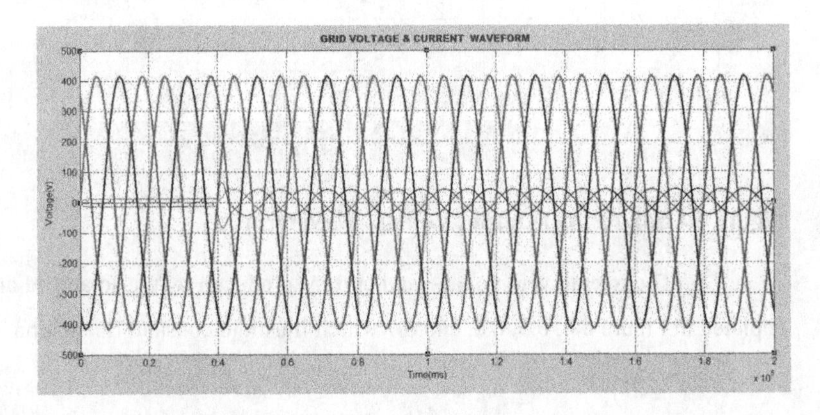

Figure 6.7 Grid Voltage and Current Waveform

The grid current and voltage show fluctuation at constant time intervals because of the variation in the power produced in single solar PV cell. The accumulated current and voltage experience constant fluctuation because of the variations in amount of irradiation incident on the PV cell. The grid current and voltage are analyzed for the THD and harmonic distortion of the voltage produced by the PV cells is measured and the pictorial representation is shown in Figure 6.8.

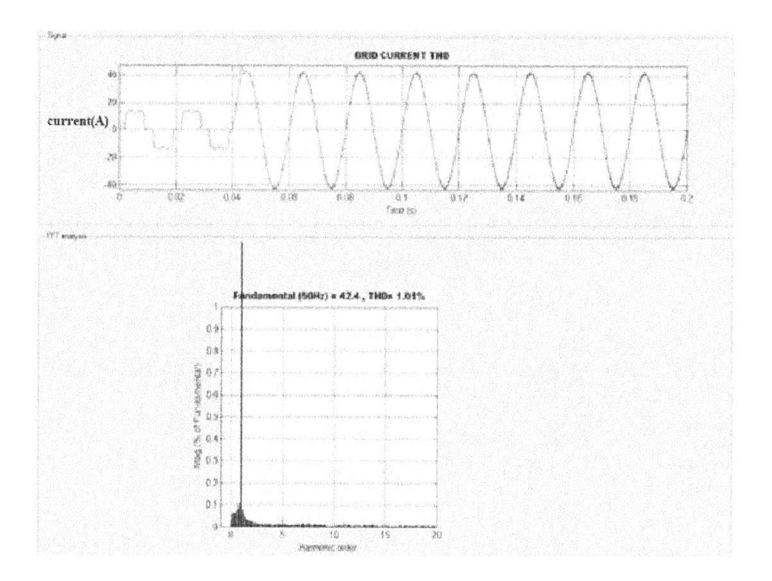

Figure 6.8 Total Harmonic Distortion (THD) in solar grid current

The THD in the solar grid voltage is evaluated by the present module of solar grids and it is determined as the additive ratio of harmonic component to a fundamental frequency it is regarded as the presented method is 50Hz and the measured THD is 1.01%.

The THD is decreased by fuzzy based MPPT controlling system and the accuracy is contrasted to the traditional PI controlling system. The presented fuzzy based MPPT controlling system is very effective when compared to the present PI controlling system and the waveform and FFT demonstration of existing PI controlling system and fuzzy based control system is represented in Figure 6.9 and 6.10.

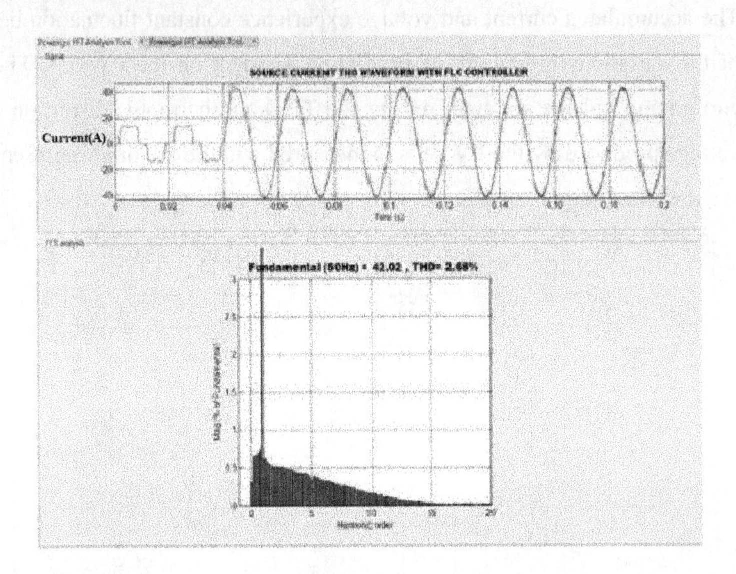

Figure 6.9 Solar Current THD waveform with fuzzy logic controller

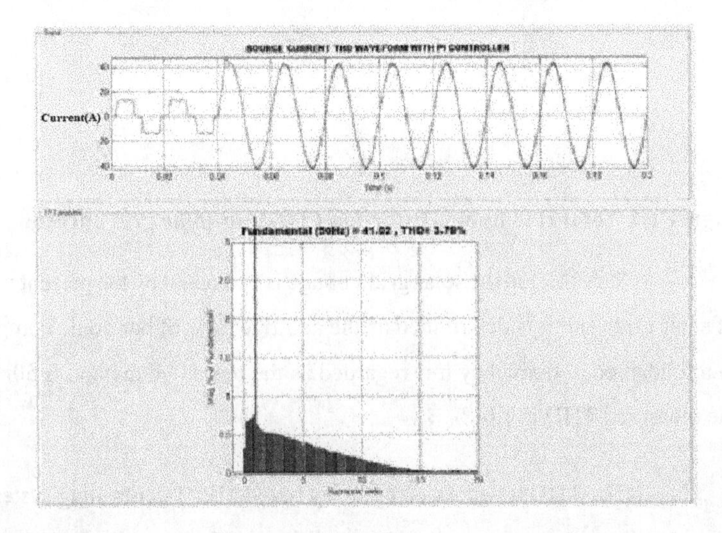

Figure 6.10 Solar Current THD waveform with PI controller

The assessment of the PI and fuzzy based controlling system is portrayed in previously described figure produces that the presented fuzzy controller poses a THD of 2.68% whereas the current PI controller shows a THD of 3.79%.

The suggested methodology comprised of five levels of processing and after that the fuzzy based controlling system, the output current is portrayed in Figure 6.11.

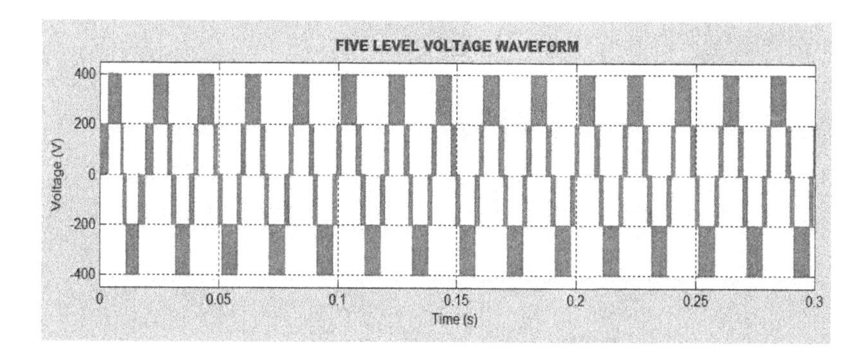

Figure 6.11 Output Voltage Waveform

Figure 6.11 illustrates the output current waveform with equal fluctuation amongst each solar grid power panel.

6.4 RESULT ANALYSIS OF FSPOA TECHNIQUE

In MATLAB, the experiment is realized using the solar PV module. The Solar cell is set up to match the parameters of the module testing facility. The results obtained were compared to other current approaches to demonstrate the suggested approach's outstanding quality. In this work, the grid current, source current, and solar current are described in Figures 6.12 to 6.16.

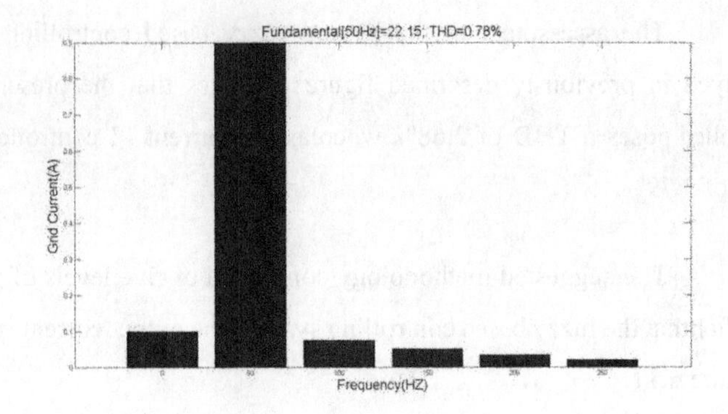

Figure 6.12 Grid current (A)

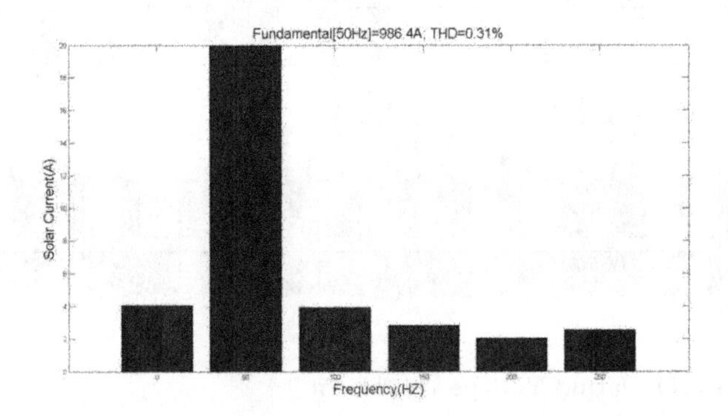

Figure 6.13 Solar current (A)

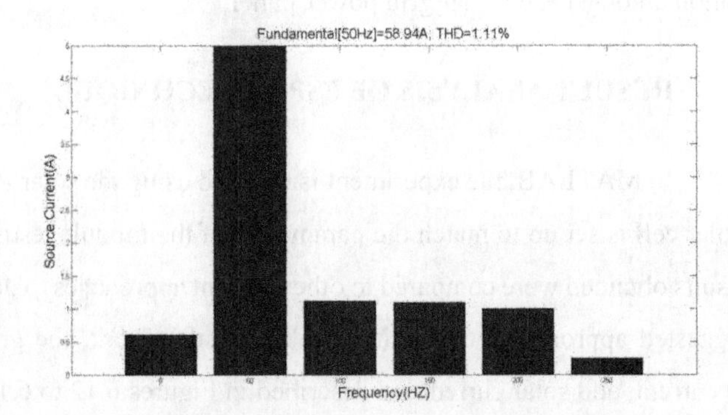

Figure 6.14 Source current (A)

By using FSPOA, the lower order harmonics can be neglected invariant levels of frequencies. The main goal of this suggested work is to eliminate the lower-order harmonics such as third-order harmonics. Besides the fundamental frequency component, other frequency components have lower-order harmonics that are illustrated. Tracking efficiency of the suggested work compared with the existing works is represented in Figure 6.15. Here, the higher efficacy of the suggested work is proved in this work. Similarly, the tracking speed of the suggested work compared with the existing works is illustrated in Figure 6.16.

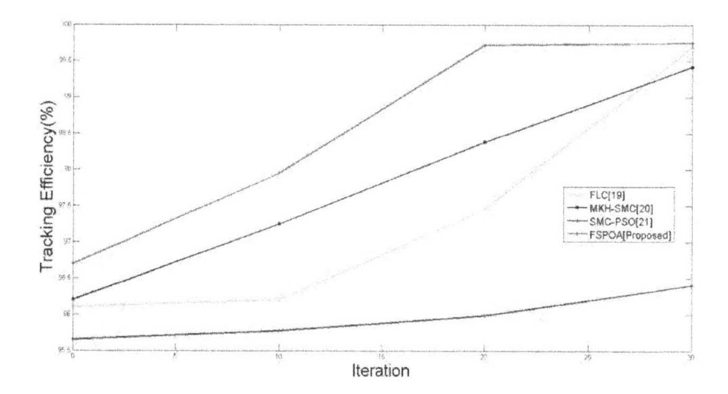

Figure 6.15 Tracking efficiency

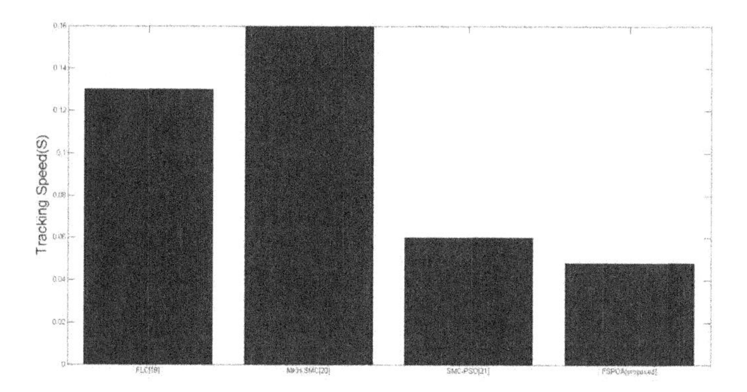

Figure 6.16 Tracking speed (s)

The ratio of the algebraic sum of all harmonic components to the power of the fundamental frequency is the total harmonic distortion, which is an assessment of the harmonic distortion present in the waveform. The formulation of THD is denoted by below two Equations (6.4) & (6.5). Figure 6.17 depicts the THD in percentage for various orders of harmonics that leads to the most mitigated percentage of THD in this work.

$$I_{THD} = \frac{\sqrt{\sum_{i=2}^{n} {I_{i rms}}^2}}{I_{1rms}} \tag{6.4}$$

Similarly,

$$V_{THD} = \frac{\sqrt{\sum_{i=2}^{n} {V_{i rms}}^2}}{V_{1rms}} \tag{6.5}$$

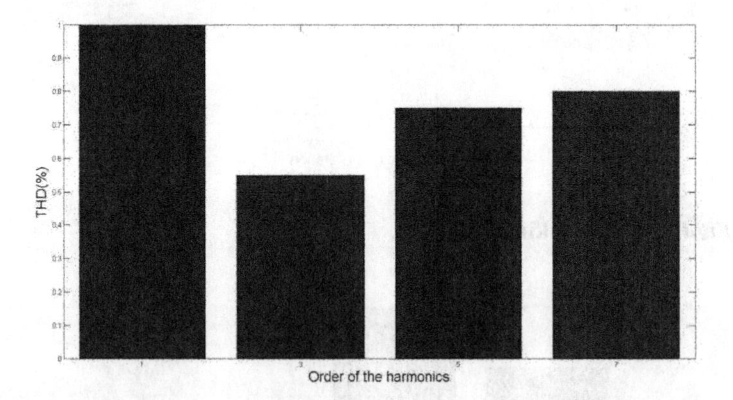

Figure 6.17 THD analysis of FSPOA (%)

Grid Parameters

Rated Power: 16 KW

Rated Voltage: 400 V

Grid Current: 40 Amps

Grid Frequency: 50 HZ

6.5 SUMMARY

This chapter has carried out a comprehensive performance that focuses on the detailed experimental validation of the three proposed models. The entire chapter is organized into three major parts. In section 6.2, the presented technique of predicting the solar irradiation in the solar PV cells utilizing KNN technique of AI was planned and implemented utilizing MATLAB 2021a. Afterward, in section 6.3, the experimental outcomes for design of fuzzy based MPPT controller with the examination of several waveforms are offered in detail. At last, in section 6.4, a detailed outcome analysis of FSPOA approach was executed utilizing MATLAB tool and the outcomes were widely examined.

CHAPTER 7

CONCLUSION AND FUTURE WORK

7.1 CONCLUSION

Since solar power generation is based on the amount of solar radiation, solar energy prediction is treated as an effective power generation scheme for effective load management. Few well known solar power forecasting models are statistical models, ML, FL, and hybrid models. The solar PV system can be combined with the SG via PLL, string inverter, micro-inverter, etc. Several challenging issues exist such as an expensive PLL-based clock driver. In this view, this research work has aimed to design PV system with optimal solar power production forecasting and MPPT maximization system. The first objective has developed K-Nearest KNN based solar radiation prediction which is analyzed quantitatively and is introduced. To ensure the goodness of the proposed models, a detailed experimental analysis is carried out using MATLAB 2021a tool. The suggested algorithm outperformed well when compared to the present models and provide higher level of accuracy with RMSE of 0.576% and consequently provide an accuracy of more than 99.424% which is regarded to be the more accurate and efficient forecasting methodologies. The effective SPP+ not only enhances the capability of gird connection and security, it also achieves minimal light discarding.

In the next objective, a fuzzy based MPPT controlling system is presented and is quantitatively analyzed. The experimental validation of proposed model is carried out for 6 months with predefined datasets and used

IEEE bus systems. In addition, the fuzzy based MPPT controlling system is investigated with present PI controlling system for the critical parameter of THD. The presented fuzzy based MPPT controller obtains an effective THD of 1.01% whereas the THD of the PI controller is 3.79%.

Finally, the third objective has presented a novel FSPOA to enhance the optimization of the MPPPT converter. Also, the voltage can be synchronized by utilizing the 3-phase transformer which means the parallel operation was performed. In this study, we get accurate output that is nearly sine waveform. The comparison of THD represents fully reduced harmonics in the output waveform. Besides, the proposed FSPOA technique shows promising performance over the other models with fully reduced harmonics in the output waveform. Therefore, it is evident that the proposed models have shown effective results over other models on PV power generation systems.

7.2 FUTURE WORK

In future, hybrid optimization algorithms can be used to improve the voltage profile and reduce the power loss of the FLCs. Besides, effective pre-processing approaches for experimental data need to be designed to summarize and extract information data. Besides, it is needed to develop new metaheuristic algorithms for training the AI models to accomplish enhanced performance. As a part of future scope, regional forecasting for energy dispatching need to be investigated. Finally, several studies have considered cloud cover as a meteorological factor and the partial shading of the PV panel is not considered, which can result in multi-peak phenomenon of the PV curve. This problem will be investigated in the future for robust prediction performance.

REFERENCES

1. Abadi, I, Imron, C & Noriyati, RD 2018, 'Implementation of maximum power point tracking (MPPT) technique on solar tracking system based on adaptive neuro-fuzzy inference system (ANFIS)', In E3s web of conferences, EDP Sciences, vol. 43, p. 01014.

2. Ab-BelKhair, A, Rahebi, J & Abdulhamed Mohamed Nureddin, A, 2020, 'A study of deep neural network controller-based power quality improvement of hybrid PV/Wind systems by using smart inverter', International Journal of Photoenergy.

3. Ahmed, A & Khalid, M 2019, 'A review on the selected applications of forecasting models in renewable power systems', Renewable and Sustainable Energy Reviews, vol. 100, pp.9-21.

4. Almeida, MP, Perpinan, O & Narvarte, L 2015, 'PV power forecast using a nonparametric PV model', Solar Energy, vol. 115, pp.354-368.

5. Antonanzas-Torres, F, Antonanzas, J, Urraca, R, Alia-Martinez, M & Martinez-de-Pison, FJ 2016, 'Impact of atmospheric components on solar clear-sky models at different elevation: Case study Canary Islands', Energy Conversion and Management, vol. 109, pp.122-129.

6. Aouchiche, N, Aitcheikh, MS, Becherif, M & Ebrahim, MA 2018, 'AI-based global MPPT for partial shaded grid connected PV plant via MFO approach', Solar Energy, vol. 171, pp.593-603.

7. Aybar-Ruiz, A, Jiménez-Fernández, S, Cornejo-Bueno, L, Casanova-Mateo, C, Sanz-Justo, J, Salvador-González, P & Salcedo-Sanz, S, 2016, 'A novel grouping genetic algorithm–extreme learning machine approach for global solar radiation prediction from numerical weather models inputs', Solar Energy, vol. 132, pp.129-142.

8. Azimi, R, Ghayekhloo, M & Ghofrani, M 2016, 'A hybrid method based on a new clustering technique and multilayer perceptron neural networks for hourly solar radiation forecasting', Energy Conversion and Management, vol. 118, pp.331-344.

9. Babar, B, Luppino, LT, Boström, T & Anfinsen, SN, 2020, 'Random forest regression for improved mapping of solar irradiance at high latitudes', Solar Energy, vol. 198, pp.81-92.

10. Badescu, V, Gueymard, CA, Cheval, S, Oprea, C, Baciu, M, Dumitrescu, A, Iacobescu, F, Milos, I & Rada, C 2013, 'Accuracy analysis for fifty-four clear-sky solar radiation models using routine hourly global irradiance measurements in Romania', Renewable Energy, vol. 55, pp.85-103.

11. Bakker, K, Whan, K, Knap, W & Schmeits, M 2019, 'Comparison of statistical post-processing methods for probabilistic NWP forecasts of solar radiation', Solar Energy, vol. 191, pp.138-150.

12. Bana, PR, Panda, KP, Padmanaban, S, Mihet-Popa, L, Panda, G & Wu, J 2020, 'Closed-loop control and performance evaluation of reduced part count multilevel inverter interfacing grid-connected PV system', IEEE Access, vol. 8, pp.75691-75701.

13. Baser, F & Demirhan, H 2017, 'A fuzzy regression with support vector machine approach to the estimation of horizontal global solar radiation', Energy, vol. 123, pp.229-240.

14. Behera, TK, Behera, MK & Nayak, N 2018, 'Spider monkey based improve P&O MPPT controller for photovoltaic generation system', In 2018 Technologies for Smart-City Energy Security and Power (ICSESP), IEEE, pp. 1-6 [March 2018].

15. Belaid, S & Mellit, A 2016, 'Prediction of daily and mean monthly global solar radiation using support vector machine in an arid climate', Energy Conversion and Management, vol. 118, pp.105-118.

16. Benali, L, Notton, G, Fouilloy, A, Voyant, C & Dizene, R 2019, 'Solar radiation forecasting using artificial neural network and random forest methods: Application to normal beam, horizontal diffuse and global components', Renewable energy, vol. 132, pp.871-884.

17. Benghanem, M & Mellit, A 2010, 'Radial Basis Function Network-based prediction of global solar radiation data: Application for sizing of a stand-alone photovoltaic system at Al-Madinah, Saudi Arabia', Energy, vol. 35, no. 9, pp.3751-3762.

18. Bimenyimana, S, Wang, C, Nduwamungu, A, Asemota, G.N.O, Utetiwabo, W, Ho, CL, Niyonteze, JDD, Hagumimana, N, Habineza, T, Bashir, W & Mesa, CK 2021, 'Integration of Microgrids and Electric Vehicle Technologies in the National Grid as the Key Enabler to the Sustainable Development for Rwanda', International Journal of Photoenergy.

19. Bouzerdoum, M, Mellit, A & Pavan, AM 2013, 'A hybrid model (SARIMA–SVM) for short-term power forecasting of a small-scale grid-connected photovoltaic plant', Solar energy, vol. 98, pp.226-235.

20. Bouzgou, H & Gueymard, CA 2017, 'Minimum redundancy–maximum relevance with extreme learning machines for global solar radiation forecasting: Toward an optimized dimensionality reduction for solar time series', Solar Energy, vol. 158, pp.595-609.

21. Box, GE, Jenkins, GM, Reinsel, GC & Ljung, GM 2015, 'Time series analysis: forecasting and control', John Wiley & Sons.

22. Campbell, SD & Diebold, FX 2005, 'Weather forecasting for weather derivatives', Journal of the American Statistical Association, vol. 100, no. 469, pp.6-16.

23. Chang, GW & Lu, HJ 2018, 'Integrating gray data preprocessor and deep belief network for day-ahead PV power output forecast', IEEE Transactions on Sustainable Energy, vol. 11, no. 1, pp.185-194.

24. Chen, C, Duan, S, Cai, T & Liu, B 2011, 'Online 24-h solar power forecasting based on weather type classification using artificial neural network', Solar Energy, vol. 85, no. 11, pp.2856-2870.

25. Cheng, HY, 2016, 'Hybrid solar irradiance now-casting by fusing Kalman filter and regressor', Renewable Energy, vol. 91, pp.434-441.

26. Chow, CW, Urquhart, B, Lave, M, Dominguez, A, Kleissl, J, Shields, J & Washom, B 2011, 'Intra-hour forecasting with a total sky imager at the UC San Diego solar energy testbed', Solar Energy, vol. 85, no. 11, pp.2881-2893.

27. Coimbra, CF, Pedro, HT & Kleissl, J 2013, 'Stochastic learning methods', Solar energy forecasting and resource assessment, pp.383-406.

28. Cornejo-Bueno, L, Casanova-Mateo, C, Sanz-Justo, J & Salcedo-Sanz, S 2019, 'Machine learning regressors for solar radiation estimation from satellite data', Solar Energy, vol. 183, pp.768-775.

29. Cui, C, Zou, Y, Wei, L & Wang, Y 2019, 'Evaluating combination models of solar irradiance on inclined surfaces and forecasting photovoltaic power generation', IET Smart Grid, vol. 2, no. 1, pp.123-130.

30. Dehghani, M, Taghipour, M, Gharehpetian, GB & Abedi, M 2020, 'Optimized fuzzy controller for MPPT of grid-connected PV systems in rapidly changing atmospheric conditions', Journal of Modern Power Systems and Clean Energy, vol. 9, no. 2, pp.376-383.

31. Dong, J, Olama, MM, Kuruganti, T, Melin, AM, Djouadi, SM, Zhang, Y & Xue, Y 2020, 'Novel stochastic methods to predict short-term solar radiation and photovoltaic power', Renewable Energy, vol. 145, pp.333-346.

32. Dowell, J & Pinson, P 2015, 'Very-short-term probabilistic wind power forecasts by sparse vector autoregression', IEEE Transactions on Smart Grid, vol. 7, no. 2, pp.763-770.

33. Duman, S, Yorukeren, N & Altas, IH 2018, 'A novel MPPT algorithm based on optimized artificial neural network by using FPSOGSA for standalone photovoltaic energy systems', Neural Computing and Applications, vol. 29, no. 1, pp.257-278.

34. Durrani, SP, Balluff, S, Wurzer, L & Krauter, S 2018, 'Photovoltaic yield prediction using an irradiance forecast model based on multiple neural networks', Journal of Modern Power Systems and Clean Energy, vol. 6, no. 2, pp.255-267.

35. El-Baz, W, Tzscheutschler, P & Wagner, U 2018, 'Day-ahead probabilistic PV generation forecast for buildings energy management systems', Solar Energy, vol. 171, pp.478-490.

36. Eseye, AT, Zhang, J & Zheng, D 2018, 'Short-term photovoltaic solar power forecasting using a hybrid Wavelet-PSO-SVM model based on SCADA and Meteorological information', Renewable energy, vol. 118, pp.357-367.

37. Fan, J, Wu, L, Ma, X, Zhou, H & Zhang, F 2020, 'Hybrid support vector machines with heuristic algorithms for prediction of daily diffuse solar radiation in air-polluted regions', Renewable Energy, vol. 145, pp.2034-2045.

38. Farayola, AM, Hasan, AN & Ali, A 2018, 'Optimization of PV systems using data mining and regression learner MPPT techniques', Indonesian Journal of Electrical Engineering and Computer Science, vol. 10, no. 3, pp.1080-1089.

39. Feng, C & Zhang, J 2020, 'SolarNet: A sky image-based deep convolutional neural network for intra-hour solar forecasting', Solar Energy, vol. 204, pp.71-78.

40. Feng, Y, Hao, W, Li, H, Cui, N, Gong, D & Gao, L, 2020, 'Machine learning models to quantify and map daily global solar radiation and photovoltaic power', Renewable and Sustainable Energy Reviews, vol. 118, p.109393.

41. Gao, B, Huang, X, Shi, J, Tai, Y & Zhang, J 2020, 'Hourly forecasting of solar irradiance based on CEEMDAN and multi-strategy CNN-LSTM neural networks', Renewable Energy, vol. 162, pp.1665-1683.

42. Gigoni, L, Betti, A, Crisostomi, E, Franco, A, Tucci, M, Bizzarri, F & Mucci, D 2017, 'Day-ahead hourly forecasting of power generation from photovoltaic plants', IEEE Transactions on Sustainable Energy, vol. 9, no. 2, pp.831-842.

43. Gupta, P & Singh, R 2021, 'Univariate model for hour ahead multi-step solar irradiance forecasting', In 2021 IEEE 48th Photovoltaic Specialists Conference (PVSC), IEEE, pp. 0494-0501 [June 2021].

44. Gutierrez-Corea, FV, Manso-Callejo, MA, Moreno-Regidor, MP & Manrique-Sancho, MT 2016, 'Forecasting short-term solar irradiance based on artificial neural networks and data from neighboring meteorological stations', Solar Energy, vol. 134, pp.119-131.

45. Haque, AU, Nehrir, MH & Mandal, P 2013, 'Solar PV power generation forecast using a hybrid intelligent approach', In 2013 IEEE Power & Energy Society General Meeting IEEE, pp. 1-5, [July 2013].

46. Hassan, MA, Khalil, A, Kaseb, S & Kassem, MA, 2017, 'Exploring the potential of tree-based ensemble methods in solar radiation modeling', Applied Energy, vol. 203, pp.897-916.

47. Hossain, M, Mekhilef, S, Danesh, M, Olatomiwa, L & Shamshirband, S, 2017, 'Application of extreme learning machine for short term output power forecasting of three grid-connected PV systems', Journal of Cleaner Production, vol. 167, pp.395-405.

48. Huang, J, Khan, MM, Qin, Y & West, S, 2019, 'Hybrid Intra-hour Solar PV Power Forecasting using Statistical and Skycam-based Methods', In 2019 IEEE 46th Photovoltaic Specialists Conference (PVSC), IEEE pp. 2434-2439, [June 2019]

49. Huang, YP, Chen, X & Ye, CE, 2018, 'A hybrid maximum power point tracking approach for photovoltaic systems under partial shading conditions using a modified genetic algorithm and the firefly algorithm', International Journal of Photoenergy.

50. Ibrahim, IA & Khatib, T 2017, 'A novel hybrid model for hourly global solar radiation prediction using random forests technique and firefly algorithm', Energy Conversion and Management, vol. 138, pp.413-425.

51. Iftikhar, R, Ahmad, I, Arsalan, M, Naz, N, Ali, N & Armghan, H, 2018, 'MPPT for photovoltaic system using nonlinear controller', International Journal of Photoenergy.

52. Jain, A, Joshi, K, Behal, A & Mohan, N 2006, 'Voltage regulation with STATCOMs: modeling, control and results', IEEE Transactions on Power delivery, vol. 21, no. 2, pp.726-735.

53. Jallal, MA, Chabaa, S & Zeroual, A, 2020, 'A novel deep neural network based on randomly occurring distributed delayed PSO algorithm for monitoring the energy produced by four dual-axis solar trackers', Renewable Energy, vol. 149, pp.1182-1196.

54. Jang, HS, Bae, KY, Park, HS & Sung, DK, 2016, 'Solar power prediction based on satellite images and support vector machine'', IEEE Transactions on Sustainable Energy, vol. 7, no. 3, pp.1255-1263.

55. Kapić, A, Zečević, Ž & Krstajić, B 2018, 'An efficient MPPT algorithm for PV modules under partial shading and sudden change in irradiance', In 2018 23rd International Scientific-Professional Conference on Information Technology (IT), IEEE, pp. 1-4, [February 2018].

56. Kaur, D, Aujla, G.S, Kumar, N, Zomaya, A.Y, Perera, C & Ranjan, R 2018, 'Tensor-based big data management scheme for dimensionality reduction problem in smart grid systems: SDN perspective', IEEE Transactions on Knowledge and Data Engineering, vol. 30, no. 10, pp.1985-1998.

57. Kaur, D, Islam, SN & Mahmud, MA, 2021, 'A Variational Autoencoder-Based Dimensionality Reduction Technique for Generation Forecasting in Cyber-Physical Smart Grids', In 2021 IEEE International Conference on Communications Workshops (ICC Workshops), IEEE, pp. 1-6, [June 2021].

58. Kenjrawy, H, Makdisie, C, Houssamo, I & Mohammed, N 2022, 'New Modulation Technique in Smart Grid Interfaced Multilevel UPQC-PV Controlled via Fuzzy Logic Controller', Electronics, vol. 11, no. 6, p.919.

59. Keyrouz, F, 2018, 'Enhanced Bayesian based MPPT controller for PV systems', IEEE Power and Energy Technology Systems Journal, vol. 5, no. 1, pp.11-17.

60. Khahro, SF, Tabbassum, K, Talpur, S, Alvi, MB, Liao, X & Dong, L, 2015, 'Evaluation of solar energy resources by establishing empirical models for diffuse solar radiation on tilted surface and analysis for optimum tilt angle for a prospective location in southern region of Sindh, Pakistan', International Journal of Electrical Power & Energy Systems, vol. 64, pp.1073-1080.

61. Khosravi, A, Koury, RNN, Machado, L & Pabon, JJG, 2018, 'Prediction of hourly solar radiation in Abu Musa Island using machine learning algorithms', Journal of Cleaner Production, vol. 176, pp.63-75.

62. Krishnan GS, Kinattingal, S, Simon, SP & Nayak, PSR 2020, 'MPPT in PV systems using ant colony optimisation with dwindling population', IET Renewable Power Generation, vol. 14, no. 7, pp.1105-1112.

63. Kroposki, B, 2017, 'Integrating high levels of variable renewable energy into electric power systems', Journal of Modern Power Systems and Clean Energy, vol. 5, no. 6, pp.831-837.

64. Kumar, N, Singh, B & Panigrahi, BK 2019, 'LLMLF-based control approach and LPO MPPT technique for improving performance of a multifunctional three-phase two-stage grid integrated PV system', IEEE Transactions on Sustainable Energy, vol. 11, no. 1, pp.371-380.

65. Kumar, N, Singh, B & Panigrahi, BK, 2019, 'Integration of solar PV with low-voltage weak grid system: Using maximize-M Kalman filter and self-tuned P&O algorithm', IEEE Transactions on Industrial Electronics, vol. 66, no. 11, pp.9013-9022.

66. Laggoun, ZEZ, Khalile, N & Benalla, H 2019, 'A Comparative study between DPC-SVM and PDPC-SVM', In 2019 International Conference on Advanced Electrical Engineering (ICAEE), IEEE, pp. 1-5, [November 2019].

67. Lalith Pankaj Raj, GN & Kirubakaran, V 2021, 'Energy efficiency enhancement and climate change mitigations of SMEs through grid-interactive solar photovoltaic system', International Journal of Photoenergy.

68. Laveyne, JI, Bozalakov, D, Van Eetvelde, G & Vandevelde, L, 2020, 'Impact of solar panel orientation on the integration of solar energy in low-voltage distribution grids', International Journal of Photoenergy.

69. Lee, W, Kim, K, Park, J, Kim, J & Kim, Y 2018, 'Forecasting solar power using long-short term memory and convolutional neural networks', IEEE Access, vol. 6, pp.73068-73080.

70. Li, H, Yang, D, Su, W, Lü, J & Yu, X 2018, 'An overall distribution particle swarm optimization MPPT algorithm for photovoltaic system under partial shading', IEEE Transactions on Industrial Electronics, vol. 66, no. 1, pp.265-275.

71. Li, Y, Su, Y & Shu, L 2014, 'An ARMAX model for forecasting the power output of a grid connected photovoltaic system', Renewable Energy, vol. 66, pp.78-89.

72. Liu, J, Fang, W, Zhang, X & Yang, C 2015, 'An improved photovoltaic power forecasting model with the assistance of aerosol index data', IEEE Transactions on Sustainable Energy, vol. 6, no. 2, pp.434-442.

73. Liu, Y, Qin, H, Zhang, Z, Pei, S, Wang, C, Yu, X, Jiang, Z & Zhou, J 2019, 'Ensemble spatiotemporal forecasting of solar irradiation using variational Bayesian convolutional gate recurrent unit network', Applied Energy, vol. 253, p.113596.

74. Macaulay, J & Zhou, Z 2018, 'A fuzzy logical-based variable step size P&O MPPT algorithm for photovoltaic system', Energies, vol. 11, no. 6, p.1340.

75. Maheshwari, NI & Chandrasekaran, M 2019, 'Harmonic analysis of photovoltaic-fed symmetric multilevel inverter using modified artificial neural network', Applied Mathematics and Information Sciences, vol. 13, no. 1, pp.105-113.

76. Mao, M, Duan, Q, Duan, P & Hu, B 2018, 'Comprehensive improvement of artificial fish swarm algorithm for global MPPT in PV system under partial shading conditions', Transactions of the Institute of Measurement and Control, vol. 40, no. 7, pp.2178-2199.

77. Mathiesen, P & Kleissl, J 2011, 'Evaluation of numerical weather prediction for intra-day solar forecasting in the continental United States', Solar Energy, vol. 85, no. 5, pp.967-977.

78. Mellit, A, Pavan, AM & Lughi, V 2014, 'Short-term forecasting of power production in a large-scale photovoltaic plant', Solar Energy, vol. 105, pp.401-413.

79. Mesloub, A, Ghosh, A, Albaqawy, GA, Noaime, E & Alsolami, BM 2020, 'Energy and daylighting evaluation of integrated semitransparent photovoltaic windows with internal light shelves in open-office buildings', Advances in Civil Engineering.

80. Moghaddamnia, A, Remesan, R, Kashani, M.H, Mohammadi, M, Han, D & Piri, J 2009, 'Comparison of LLR, MLP, Elman, NNARX and ANFIS Models - with a case study in solar radiation estimation', Journal of Atmospheric and Solar-Terrestrial Physics, vol. 71, no. 8-9, pp.975-982.

81. Monteiro, V, Sousa, TJ, Sepulveda, MJ, Couto, C, Martins, JS & Afonso, JL 2019, 'A novel multilevel converter for on-grid interface of renewable energy sources in smart grids', In 2019 international conference on smart energy systems and technologies (SEST) IEEE, pp. 1-6, [September 2019].

82. Monteiro, VDF, Sousa, TJ, Pinto, JG & Afonso, JL, 2018, 'A novel front-end multilevel converter for renewable energy systems in smart grids'.

83. Motaparthi, N & Malligunta, KK 2022, 'Seven level aligned multilevel inverter with new SPWM technique for PV, wind, battery-based hybrid standalone system', International Journal of Emerging Electric Power Systems.

84. Mukundan, NMC, Pychadathil, J, Subramaniam, U & Almakhles, DJ 2020, 'Trinary hybrid cascaded H-bridge multilevel inverter-based grid-connected solar power transfer system supporting critical load', IEEE Systems Journal, vol. 15, no. 3, pp.4116-4125.

85. Nabae, A, Takahashi, I & Akagi, H 1981, 'A new neutral-point-clamped PWM inverter', IEEE Transactions on industry applications, vol. 5, pp.518-523.

86. Nasir, T, Bukhari, SSH, Raza, S, Munir, HM, Abrar, M, Bhatti, KL, Ro, JS & Masroor, R 2021, 'Recent Challenges and Methodologies in Smart Grid Demand Side Management: State-of-the-Art Literature Review', Mathematical Problems in Engineering.

87. Nerubatskyi, V, Plakhtii, O, Hordiienko, D & Khoruzhevskyi, H 2020, 'Study of energy parameters in alternative power source microgrid systems with multilevel inverters', Industry 4.0, vol. 5, no. 3, pp.118-121.

88. Nureddin, AAM, Rahebi, J & Ab-BelKhair, A, 2020, 'Power management controller for microgrid integration of hybrid PV/fuel cell system based on artificial deep neural network', International Journal of Photoenergy.

89. Panda, DK & Das, S, 2021, 'Smart grid architecture model for control, optimization and data analytics of future power networks with more renewable energy', Journal of Cleaner Production, vol. 301, p.126877.

90. Pedro, HT & Coimbra, CF 2012, 'Assessment of forecasting techniques for solar power production with no exogenous inputs', Solar Energy, vol. 86, no. 7, pp.2017-2028.

91. Perez, R, Kivalov, S, Schlemmer, J, Hemker Jr, K, Renné, D & Hoff, TE 2010, 'Validation of short and medium term operational solar radiation forecasts in the US', Solar Energy, vol. 84, no. 12, pp.2161-2172.

92. Peters, IM & Buonassisi, T 2019, 'The impact of global warming on silicon PV energy yield in 2100', In 2019 IEEE 46th Photovoltaic Specialists Conference (PVSC), IEEE, pp. 3179-3181, [June 2019].

93. Pires, VF, Cordeiro, A, Foito, D & Silva, JF, 2018, 'Three-phase multilevel inverter for grid-connected distributed photovoltaic systems based in three three-phase two-level inverters', Solar Energy, vol. 174, pp.1026-1034.

94. Prabaharan, N & Palanisamy, K 2016, 'Analysis and integration of multilevel inverter configuration with boost converters in a photovoltaic system', Energy Conversion and Management, vol. 128, pp.327-342.

95. Priyadarshi, N, Padmanaban, S, Holm-Nielsen, JB, Blaabjerg, F & Bhaskar, MS 2019, 'An experimental estimation of hybrid ANFIS–PSO-based MPPT for PV grid integration under fluctuating sun irradiance', IEEE Systems Journal, vol. 14, no. 1, pp.1218-1229.

96. Quan, H, Khosravi, A, Yang, D & Srinivasan, D 2019, 'A survey of computational intelligence techniques for wind power uncertainty quantification in smart grids', IEEE Transactions on Neural Networks and Learning Systems, vol. 31, no. 11, pp.4582-4599.

97. Rana, M, Koprinska, I & Agelidis, VG, 2016, 'Univariate and multivariate methods for very short-term solar photovoltaic power forecasting', Energy Conversion and Management, vol. 121, pp.380-390.

98. Raza, MQ, Nadarajah, M & Ekanayake, C, 2016, 'On recent advances in PV output power forecast', Solar Energy, vol. 136, pp.125-144.

99. Reikard, G, 2009, 'Predicting solar radiation at high resolutions: A comparison of time series forecasts', Solar energy, vol. 83, no. 3, pp.342-349.

100. Ren, Y, Suganthan, PN & Srikanth, N, 2015. 'Ensemble methods for wind and solar power forecasting - A state-of-the-art review', Renewable and Sustainable Energy Reviews, vol. 50, pp.82-91.

101. Selvaraj, DA & Victor, K 2021, 'Design and performance of solar PV integrated domestic vapor absorption refrigeration system', International Journal of Photoenergy.

102. Sharew, EA, Kefale, HA & Werkie, YG 2021, 'Power Quality and Performance Analysis of Grid-Connected Solar PV System Based on Recent Grid Integration Requirements', International Journal of Photoenergy.

103. Sharma, B, Dahiya, R & Nakka, J 2019, 'Effective grid connected power injection scheme using multilevel inverter based hybrid wind solar energy conversion system', Electric Power Systems Research, vol. 171, pp.1-14.

104. Sheng, H, Xiao, J, Cheng, Y, Ni, Q & Wang, S 2017, 'Short-term solar power forecasting based on weighted Gaussian process regression', IEEE Transactions on Industrial Electronics, vol. 65, no. 1, pp.300-308.

105. Song, Y, Wu, D, Wagdy Mohamed, A, Zhou, X, Zhang, B & Deng, W 2021, 'Enhanced success history adaptive DE for parameter optimization of photovoltaic models', Complexity.

106. Stonier, AA, Murugesan, S, Samikannu, R, Venkatachary, SK, Kumar, SS & Arumugam, P 2020, 'Power quality improvement in solar fed cascaded multilevel inverter with output voltage regulation techniques', IEEE Access, 8, pp.178360-178371.

107. Sun, S & Huang, R 2010, 'An adaptive k-nearest neighbor algorithm', In 2010 seventh international conference on fuzzy systems and knowledge discovery, IEEE, vol. 1, pp. 91-94, [August 2010].

108. Suvetha, PS & Seyezhai, R 2022, 'Design and Development of PV/FC Based Integrated Multilevel Inverter for Smart Grid', In ISUW 2019, Springer, Singapore, pp. 167-177.

109. Syahputra, R & Soesanti, I 2020, 'Planning of hybrid micro-hydro and solar photovoltaic systems for rural areas of central Java, Indonesia', Journal of Electrical and Computer Engineering.

110. Tchao, ET, Quansah, DA, Klogo, GS, Boafo-Effah, F, Kotei, S, Nartey, C & Ofosu, WK, 2021, 'On cloud-based systems and distributed platforms for smart grid integration: Challenges and prospects for Ghana's Grid Network', Scientific African, vol. 12, p.e00796.

111. Tey, KS, Mekhilef, S, Seyedmahmoudian, M, Horan, B, Oo, AT & Stojcevski, A, 2018, 'Improved differential evolution-based MPPT algorithm using SEPIC for PV systems under partial shading conditions and load variation', IEEE Transactions on Industrial Informatics, vol. 14, no. 10, pp.4322-4333.

112. Tian, AQ, Chu, SC, Pan, JS & Liang, Y, 2020, 'A novel pigeon-inspired optimization based MPPT technique for PV systems', Processes, vol. 8, no. 3, p.356.

113. Vicente, EM, Vicente, PS, Moreno, RL & Ribeiro, ER 2020, 'High-efficiency MPPT method based on irradiance and temperature measurements', IET Renewable Power Generation, vol. 14, no. 6, pp.986-995.

114. Wang, X, Cui, P, Du, Y & Yang, Y 2020, 'Variational autoencoder based fault detection and location method for power distribution network', In 2020 8th International Conference on Condition Monitoring and Diagnosis (CMD), IEEE, pp. 282-285, [October 2020].

115. Wang, Z, Wang, F & Su, S, 2011, 'Solar irradiance short-term prediction model based on BP neural network', Energy Procedia, vol. 12, pp.488-494.

116. Yun, GUAN, Zhang, B & Xinyang, YAN, 2019, 'Accurate Short-term Forecasting for Photovoltaic Power Method Using RBM Combined LSTM-RNN Structure with Weather Factors Quantification', In 2019 IEEE Sustainable Power and Energy Conference (iSPEC), IEEE, pp. 797-802, [November 2019].

117. Zhang, J, Florita, A, Hodge, BM, Lu, S, Hamann, H.F, Banunarayanan, V & Brockway, AM 2015, 'A suite of metrics for assessing the performance of solar power forecasting', Solar Energy, vol. 111, pp.157-175.

118. Zhang, Y, Xu, Y & Dong, ZY 2017, 'Robust ensemble data analytics for incomplete PMU measurements-based power system stability assessment', IEEE Transactions on Power Systems, vol. 33, no. 1, pp.1124-1126.

119. Zhu, Z, Zhou, D & Fan, Z 2016, 'Short term forecast of wind power generation based on SVM with pattern matching', In *2016* IEEE International Energy Conference (ENERGYCON), IEEE, pp. 1-6, [April, 2016].

9 781916 706354

Printed by BoD™in Norderstedt, Germany